The Backyard Stonebuilder

THE BACKYARD STONEBUILDER

Stonebuilding Projects
for the Weekend Mason

by Charles Long

Warwick Publishing Group
Toronto

Published by:
Warwick Publishing Group
Toronto, Ontario

Previously published by Summerhill Press Ltd. of Toronto, Ontario.

ISBN 0-920197-19-1

Distributed in Canada and the United States by:
Firefly Books
250 Sparks Avenue
Willowdale, Ontario M2H 2S4

Printed and bound in Canada

For Kirsten and Ewan, who treat their father's foible with irreverent good humour, and occasional help.

Contents

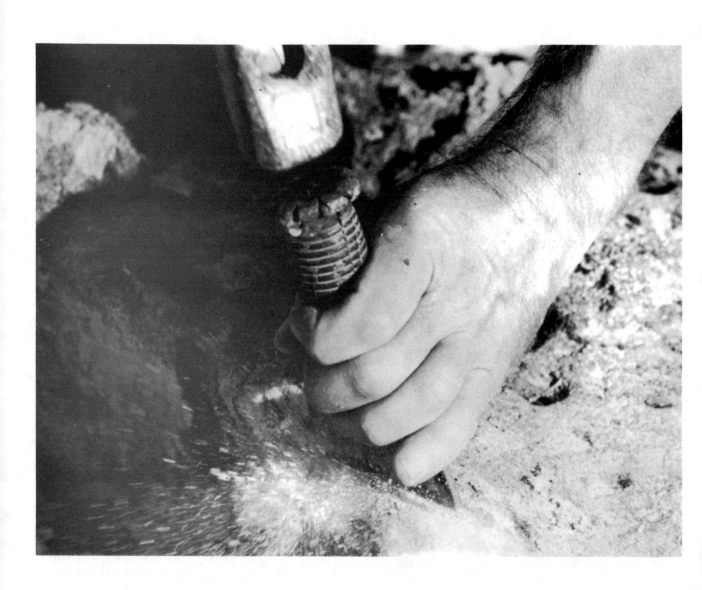

Introduction

It happens wherever bedrock makes its way to the surface. Bulldozers, clearing the way for houses, up-end chunks of stone. Frost breaks would-be building blocks out of the ledges. Rain and wind bare boulders. One way or another, some of us end up with rocks in the yard. Right out there with the dandelions and reminders of the neighbor's dog. Nature delivers.

We whipper-snip around them, roll them under the hedge, line the driveway with them. The garbage man won't take them, and the lawnmower hates them.

Sooner or later, this thought occurs to any half-handy homeowner: "Geez, I wonder if I could pile these blinkin' rocks over there and make a . . . a what? A planter? A little wall?"

At that point, most give up. Those who go ahead and try get hooked. Like kissing. Once you've tried it, there's no turning back.

Oh, there are lots of reasons not to try: it must be difficult/ these brutes are heavy/ and what if it falls over and hurts somebody. All very sensible doubts. But if you persist and actually put one rock atop another . . . and then another, don't expect to get off lightly. When the wall (or whatever) is done, you will not be able to stop yourself from touching it, the way the novice carpenter touches freshly sanded wood and marvels that he has made it feel like that. You will pat the stones and know that you have built one good thing in your life that will outlast the mortgage. And you'll want to do it again.

Like Bob Edgington. Bob retired after a long career in the public service and two heart attacks. But there was a litter of stones around the cottage, and bit-by-bit Bob got busy again. Several years later an old friend arrived and was flabbergasted by the results.

"How," asked the friend, "could a man of your age, and in your condition, build all that?"

"One stone at a time," replied Bob. And he's still doing it, thirteen years later.

Which is why the range of projects in this book is so wide: from the four-hour hearth to the Super-Hibachi, which I puttered over for weeks . . . weeks when there were a dozen more important jobs I knew I should have been doing. Some of the projects are simple, some more complex, but all of them have been built by honest-to-goodness amateurs who began by thinking it was far too difficult, too heavy, and would probably fall down away.

And after a year or so, when the whatsit still hasn't fallen down, when the neighbor has leaned over the hedge for the 93rd time muttering things like, "Sure is a beauty/ lost art, eh/ and where'd ya learn to do that anyway? . . . that's when you'll poke around under the hedge for the leftover stones, thinking of projects with a bit more challenge.

For the sake of those who do get hooked and come back for a bigger challenge (and for the sake of the printing bill), we resisted temptations to repeat basic information in every project chapter. Each chapter does include some advice on the rudiments — cutting stone, mixing mortar and that sort of thing. And the easy projects, like the hearth, include relatively more on these basics. But the more difficult projects assume that the reader is familiar with the fundamentals. By the time you start building Super-Hibachi, for example, we'll just take it for granted that you've done this kind of thing before and that further advice on over-lapping vertical joints would be a waste of ink. Anyway, the basic techniques are fully described in chapters 13 and 14. If you can't find an answer in the project chapter, look for it in the back of the book.

Naturally, the projects have more in common than the use of stone. Both the porch and the hearth, for example, involve fitting and levelling a smooth, flagstone surface. those two projects, however, are quite different in some other respects: the hearth is designed around problems of weight and bouncing joists; the porch is built to accomodate frost and drainage problems. But the fitting of a surface involves some common techniques, and readers will find it useful to browse through related chapters to see a technique applied in several circumstances.

The mudroom, on its surface, is little different from a hearth, a patio, or a porch floor; but the possibility of insulating an interior floor makes the mudroom a worthwhile project to include, if only to show off that technique.

Similarly, the smokehouse has a number of useful lessons on the effects of heat. The gatepost project illustrates the use of a "floating pad" foundation. The table involves only the most basic masonry, and yet it serves to demonstrate some very useful ideas on how to move and handle extremely heavy stones.

Whatever your interests, we hope you will be encouraged to adapt these projects to your own backyard. They aren't meant to be copied in every detail, like a blueprint. They are meant to give you some ideas. They are meant to illustrate the problems and possibilities of some different kinds of stone construction. We have tried, in other words, to concentrate on the "how-to" rather that the "what-to".

Enjoy!

Chapter 1

The Portable Hearth

Permanence, the first virtue of building in stone, may sometimes be a vice. In this yuppie age of the three year marriage and the one year mortgage, the wish to be mobile can make stone seem rather impractical for some. Like investing in thousand-year bonds. Modern furniture snaps together in portable modules. The bed rolls up, the TV has a handle on top, and the development house is built to fall apart almost as fast as the family. So why build anything that can last a thousand years? You can't take it with you. You can't even change your mind.

Well, it ain't necessarily so. Here's a real stone hearth that Randy Bolger and I built in his living room in four hours flat. It doesn't actually have a handle, but this hearth moves as fast as the Eighties. You can take it apart and pack it in the trunk of the car almost as fast as you can roll up your futon.

The story of the Bolger's woodstove will be familiar to anybody who has ever been married. To begin with, they had just installed new wall-to-wall carpet and neither Randy nor Sue could face the thought of carving out a chunk of expensive broadloom for the sake of a permanent hearth. Secondly, they're toying with the idea of selling the house and not every buyer would want a woodstove. Then there are the family allergies — any new fixture in the house may have to be pulled out quickly if Jason starts to sniffle. Finally, like any happily married couple, Randy and Sue can disagree on anything. If Randy wants bricks, Sue will prefer tile. Potato and potahto. Choosing a colour can take days. So anything new is subject to change until they've looked at it long enough to decide that they both like it.

Sue was pretty sure she wanted a wood fire before the Christmas season, but there were too many uncertainties to commit one entire corner of the living room to a thousand-year hearth. The hearth had to be safe *and* portable.

For safety's sake, the surface had to be fireproof, sparkproof, and light enough to leave the floor unbowed. For portability, the hearth had to be installed over the carpet, require no special support, and be fitted without mortar.

Frankly, there are better ways to build a flagstone hearth, but none so easy and portable as this.

Design

The hearth is there to protect the floor from sparks or embers which might escape from the firebox. Woodburners also worry about creosote fires and overheated stoves; but the basic, everyday problem is the spark that flies when you toss on a log or dig out the ashes. For that reason, building codes are primarily concerned with the areas of a hearth — its extension beyond the firedoor.

Ask your building inspector about local standards. In some jurisdictions, a special permit is required for a woodstove installation. A reputable woodstove dealer will also provide safety information for the particular model you wish to install.

Before you can lay out the hearth, you must determine where the stove will stand, how far from the wall, and how it will connect to the chimney. These are safety considerations, not aesthetic ones. As a rule of thumb, an unshielded stove should be placed at least three feet away from the nearest combustible surface. Properly installed heat shields, on the stove or on the wall, can reduce that distance, but in every case it's best to accept the advice of the building inspector and the dealer.

Many residential standards call for a hearth that extends at least 16 inches in front of the firebox opening, and at least 8 inches to either side of the opening. That's the minimum standard. Common sense would also tell us that a firebox placed high above the floor can throw a spark farther than a floor level opening; and an opening with a built-in "lip" is safer than one without.

Draw a floor plan. Better still, mark the floor itself so that you can see exactly where the stove will be. (Figure 1-1) A line drawn parallel to the front of the stove, then moved out 16 inches into the room, marks the minimum depth of the hearth. The minimum width will be the width of the firebox opening plus an extra 8 inches to either side. Now you can add to this basic shape in any way you like, for the sake of appearance or convenience. But please don't subtract anything from it.

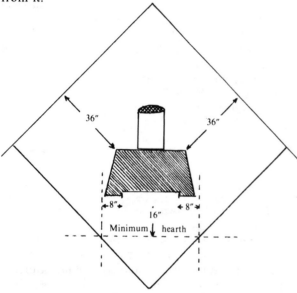

Figure 1-1   Draw a floor plan, or mark it out on the floor. The distance of the stove from the nearest wall can only be reduced if you use an approved heat shield.

The Bolger hearth was meant to be basic. Randy planned a simple square in the corner. Because it would be raised above the floor in a relatively small room, they wanted to keep it out of the way — where it would be least likely to be tripped over. Heat shields would allow them to place the stove closer to the wall.

Choose your stove, pick a location a safe distance from the walls, and plan the hearth a safe distance out from the stove door. The rest is artistic imagination and trying not to trip your friends.

The next calculation was the final thickness of the hearth. Because the Bolger hearth was to be fitted "dry" (without mortar) and placed atop the carpet, we had to design a frame that would hold the stones in place and keep the mess contained. The simple choice was a wooden box of 2 inch lumber, set on edge. The question was how high to make the sides?

As is often the case with building in stone, it is easier to amend the design than to amend the material. So it was easier to measure the stones we had available and then cut the wooden sides to fit. The stones we had selected were about 2 inches thick. We allowed another inch or so for the sand bed and any unusually fat stones. The floor of the box should be ½ inch thick. The total depth of materials, then, would be about 3½ inches. So Randy's box could be framed with ordinary 2 × 4's set on edge. Thicker stones could be accomodated by using higher sides on the box.

The limiting factor is weight. The thicker the hearth, the heavier it is. And the extra weight, at some point, will require some extra support under the floor. The bolger hearth, at 3½ inches, weights in at about 40 pounds per square foot. That's child's play for any but the flimsiest floor. Here's how to figure it:

$$\frac{\textit{thickness in inches} \times \textit{average weight per cubic foot}}{12}$$

Chances are that any stone you find that is flat enough to make a hearth will be either sandstone or limestone, the most common sedimentary rock. It is the sedimentary rock that splits easily and naturally into flat "flag" stones. Generally, sandstone will weigh a little less than 150 pounds per cubic foot, limestone a little more. Density varies, but as a rough guide assume that most flat stones will weigh about 150 pounds per cubic foot. So a 4 inch hearth would place a load on the floor of 50 pounds per square foot:

$$\frac{4 \times 150}{12} = 50 \textit{ pounds per sq foot}$$

If the surface of the 4 inch hearth is 20 square feet, the total weight would be 1000 pounds, or about the same weight as a small dinner party. Actually, the hearth puts less stain on the floor than the dinner guests, since the hearth doesn't bounce around, and it spreads its weight evenly over the surface. The comparison would be closer if you insisted that your guests wear snowshoes, or lie flat on the floor without moving.

Because of the wide distribution of the weight, Randy's stone heart won't even dent the carpet.

Materials

Flat stones
Masonry sand
2 inch lumber
Plywood
Moulding trim
Screws and nails

The starting point is the stone. First, find the stone, and then plan everything else around that. The ideal stone for this job would be flat on both sides. The actual thickness matters less than the uniformity — all the stones should be more or less the same thickness.

Randy's neighborhood is littered with fieldstone, mostly flat slabs of sandstone and limestone. Like a lot of exposed rock, however, the surfaces are badly weathered. On closer examination, the bumps and undulations spoil the smooth, flat surfaces we were looking for.

Since weathering had spoiled what had once been perfectly good surfaces, we didn't have to look far for the good stuff. In a nearby wooded area, where a carpet of leaves had cushioned the effects of rain and frost, we found just what we were looking for: thin sandstone slabs, with smooth faces and square (untapered) edges. Tree roots had broken the slabs off the surface of the bedrock, and the leaves in turn had protected the slabs from weather. Scraping around in the leaves with a small crowbar, I found an area where the slabs were a uniform 2 inches to 3 inches thick. In 15 minutes we dug up enough rock for several hearths.

How much is enough? Twice as much as you think you will need. Fitting, you'll find, is easier with a wider selection of stone from which to choose. And there will be mistakes in the cutting, turning flagstones into gravel. Finally, there is no such thing as too much stone — leftovers from one project become the genesis of the next.

If you must be scientific about quantities, lay out the stones as you gather them. Overlap the edges generously, so that the ground is completely covered. When the array is larger than the area to be covered by the hearth, then you have almost enough. Do be generous, though.

The amount of sand required depends upon the uniformity of the rock, the closeness of the fit, and the depth of the box. The sand must fill the spaces between and under the stones — the bigger the spaces, the more sand needed. Randy, still a cabinetmaker at heart, had his doubts about the fitting of stone, and so bought far more sand than we needed. In the end, we used about three or four pails, about 40 pounds.

Masonry sand is preferred, not because filling spaces is particularly demanding, but because masonry sand has been screened, and screening eliminates the larger pebbles that might interfere with a tight fit.

The springy pile of the carpet would allow the flagstones to wiggle and shift. So the box needs a stiff bottom. Plywood is ideal. 5/8ths is the minimum thickness, but make it heavier if you can afford it. If possible, use a single piece. If the hearth is so big that you can't cut the bottom from a single sheet of plywood, then do the next best thing and buy the tongue and groove sheets that are sold for rough flooring.

For the sides of the box, you will need some 2 inch lumber. The width depends on the desired depth of the box, and the total length depends on the perimeter of the hearth. Plan to continue the sides completely around the box, even where the box adjoins a wall. Naturally, the moulding or trim will be needed only along the exposed edges of the box.

Fasteners can be screws or dowels at the corners of the box, screws to hold the bottom to the sides and finishing nails for the molding. (Figure 1-2)

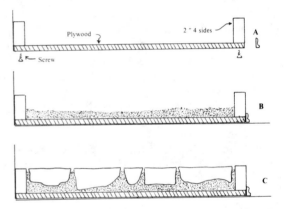

Figure 1-2 The entire project is simplified if you think of it in terms of (A) assembling the box, (B) adding sand, and (C) fitting in the stones.

Preparation

Cut the sides to length and join them at the corners. Randy used dowels and a fancy corner piece, but that has more to do with Randy than the requirements for a hearth. Attach the bottom to the sides and put the box in place. If it adjoins a wall that has been finished with baseboard and moulding, you should remove the moulding behind the box. (Photo 1-1) This is not essential, but it does eliminate a possible gap between baseboard and box — a gap that will be easier to see than to clean.

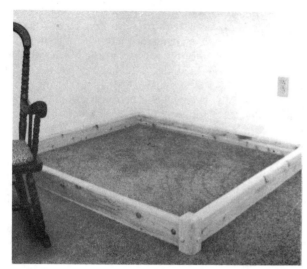

Photo 1-1 Join together the retaining box and place in position.

If the bottom of the box is rigid and tightly fastened to the sides (no cracks wide enough to leak sand), that's all the protection the carpet should need. If the bottom is not well fastened, you can line the box with 6 mil plastic. But don't take this precaution unless it is necessary. Plastic allows the sand to slide around, causing other problems.

It will not be necessary to fasten the box to the wall, or to the floor. The weight of the stones will keep everything in place.

The trim can be added when the box is in place, or, better still, leave it until last, after the stonework mess is cleared away.

And speaking of mess, Randy knew me well enough to have the entire room covered with cardboard before I arrived. The normal litter of stone chips and sand, not to mention boot tracks, could make quite a mess of the living room otherwise.

When the box is ready, dump in the sand. A bucket or two is enough to start. That's it. You're ready to start fitting the stones.

Fitting

The trick to fitting stones into a surface is to begin at the front, at the most visible, most heavily used edge. That's where the best and biggest stones should go. There are three reasons:

1) Appearance. The bigger the stone the better it looks. Small people may object, and some stones (like diamonds) look just fine, but the object there is to look

floor-like: flat, uniform, and solid. The smaller the stone, the more cracks between them, the more fragmented the look. Sorry, small people.

2) Stability. This is particularly important in mortarless construction. Small stones, set in sand, shift a little with every footstep on the surface. Even a heavy tread on a distant part of the floor can affect them. Bigger stones are less likely to shift.

3) Efficiency. No matter how carefully you measure the stones and plan, there will be narrow gaps and odd-shaped spaces at the last fitting. Better to lay the best stones first then cut the lesser rocks to fit the angles of the nice ones.

Randy's square box meant starting right in the front corner. The one, large, flat rock with a square (sort of) corner went first. (Photo 1-2) Naturally, the fit wasn't perfect. That's where the chalk and the saw come in.

Pick the best edge — the straightest, longest edge —and wiggle it up snug against the side of the box, as far into the corner as it will go. That leaves one edge fitted, and the other edge straggling off on some odd angle that may be more or less than 90°. Use the level, or a yardstick, to mark the bad edge for cutting. Line up the stick so it's parallel with the frame, and across the point where you want the corner of the stone to be. Mark the line with chalk. (Figure 1-3)

Photo 1-2 Place the first stone in the box. It should be a large piece with a squarish corner.

18

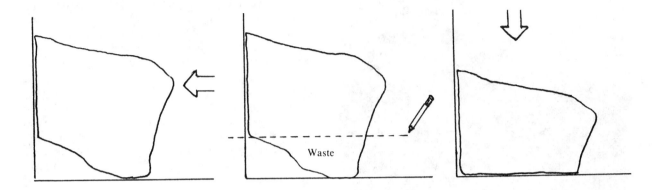

Figure 1-3   To make a square corner and ensure a nice fit, select the straightest edge and fit that edge tightly against the box (left), mark the cut parallel to the other side of the box (center), and then cut off the waste and fit the stone tightly into the corner (right).

In other kinds of stone building, the fastest way to cut is with hammer and chisel. These thin flagstones, however, are more fragile than the chunkier rock normally used in a wall. A chisel too often cracks a flagstone along some other — unintended — line. With these thin stones, the chisel must be kept sharp and applied with the lightest touch. Better still, use a "masonry" or "concrete" blade in an ordinary, electric saw.

Set the circular blade at a shallow depth — ½ inch or so — and run it lightly along the chalk line. The idea is to start the cut gradually, making it a little bit deeper on every pass. The saw wants to buck and skitter on the rough surface, and you'll have to take a firm grip on things to keep the groove running straight. Adjust the depth guide on every pass, cutting a little deeper each time. The deeper cut will eventually help to control the blade.

The dust and grit will fly. Cut outdoors, and wear a mask. Safety glasses, too. You can control some of the dust and cool things off at the same time, by keeping the cut wet. Sprinkle it, dunk it, rinse if off with every pass.

It's rarely necessary to cut all the way through. Unless the stone is unusually thin or fragile, a deep score is all you need. Cut about a third of the way through. Support the stone beneath the cut, so that the excess hangs out over the edge of the bench (or table or

whatever), and tap the excess with the hammer. Not too hard. Tap back and forth along the cut until it breaks. If you've ever cut glass, the principle is much the same.

When the rock does break along the cut, any knobs or projections that remain can be cleaned off with the hammer. Just chip them off from the good side (the scored side), and let the underside break off any way it likes. Again, be gentle with that hammer or you risk breaking the nearly finished stone across the middle.

Oddly enough, the hardest cut to make cleanly is the narrow one. When a long, thin piece has to be taken off an edge, it's prone to come away in frustrating little bits instead of one clean crack that runs the length of the saw cut. Think about that when you're planning where to cut. But if you have no choice, score more deeply than you otherwise would, and be prepared to chip away at the rough spots.

So far, we've considered only cuts that shape the surface dimension, so the edges of the stone will fit the corner, or fit the edge of the rock next door. It happens, though, that a rock may be too thick in spots to settle down to a flush surface. It "stands up" on its bumps, and can't be wriggled down in the sand any farther. If this happens with any old uninspired bit of stone, throw it out and try another. Murphy's law, however, puts the bumps on the best and biggest rocks. For them, you'll take off the bump (Photo 1-3)

With sedimentary rock, a bump will often be a bit of

Photos 1-3   *If you find that one stone is too thick and won't level with the surface, you can:*

a) pull out the stone and mark the undesireable with chalk,

b) score the bump with a series of parallel cuts to depth of the desired surface, and

c) chisel off the bump by chipping away the scored surfaces.

the next stratum that has been left behind. The bump can be "peeled" off with the chisel the way a plane takes bumps off a board. Look for a line of grain, possibly a crack, running around the base of the bump.

If the grain is more erratic, or if it's a weathered bump, it may not come away cleanly. The solution then is to chip it away in easy pieces. (Figure 1-4) First, score the bump with a series of parallel saw cuts. Each cut is slightly deeper than the bump, and the cuts should be less than an inch apart. The saw won't ride smoothly over the bump, so you'll have to raise the depth gauge and saw "freehand". The result looks like a series of ridges. Put the chisel at the bottom of the outside ridge and chip it off. That gives access to the bottom of the next ridge. Chip it off. It won't be pretty, but it works.

Figure 1-4   *When a bulge interferes with a level fit as in the drawing, you can:*

a) mark the desired depth

b) Score the bulge with saw cuts

c) Chisel off the bulge

Filling in the Frame

The first stone is in — snug against the frame, any bottom bumps chipped off. Wriggle it into the sand until the top surface is flush with the top of the frame.

Level the stone by running a straight-edge across the top of the frame. (Photo 1-4) Eye the top surface of the stone against the straight-edge. Fix the high spots by working that part deeper into the sand. Fix the low spots by raising the stone and sprinkling more sand beneath it.

When you raise the stone, you may be able to see the impression that the stone has left in the sand. The damper the sand, the clearer the impression will be. Ideally, the entire area under the rock will have been "pressed" by the stone. In practice, however, you are more likely to see shallow holes, where the stone hasn't touched the sand. Fill these in so that every part of the stone's bottom can rest solidly on the sand.

With the best and biggest stone fitted, and settled to a solid level that is flush with the top of the frame, you are ready to start filling in around it. Again, work from the front to the back. Leave the area behind the stove, or under it, until last.

There are two things to keep in mind as you pick the next rock in the pattern. The first is the angle formed by the edges — the "corner" if you like. You must find a corner that matches the angle left by the previous rock and the frame. If you can't find a matching angle, you must make one. The stone must fit the space. Sounds obvious, doesn't it?

It is obvious; and yet that simple act of matching angles — by turning the stones around and measuring them with the eye — is the heart of all masonry skills.

If no stone fits the angles of the space, then one must be cut. It is done the same way we cut the first big stone to go in the front. (Photo 1-5) Fit the best edge snug against the stone already in place. Set the straight edge parallel to the frame. Mark the cut with chalk. Score the mark with the saw. Break off the waste with a hammer. Push the cut stone into place and nestle it down into the sand until the top is flush with the frame.

Matching angles is the first consideration. The second is to keep an eye on the new edge you're forming at the back of the row. So far, we've been working against the frame as our beginning edge. All we have to do is match it with a reasonably straight edge on a rock. It doesn't matter how long the rock's straight edge might be, it will fit somewhere along the frame. But in the second and subsequent rows, we'll be working against the uneven backs of the first row of rocks. The more we can straighten out that back edge, the easier the subsequent fitting will be.

As you fill in the frame, getting closer to the last open space in the mosaic, one more truth will raise its ugly head. As long as we were filling in a row at a time,

Photo 1-4   Level the stone by running a straight edge across the top of the frame.

Photos 1-5   If a stone doesn't fit well and needs to be cut as in the stone to the left of the chalk box (top), place the stone over the adjoining stone and mark with a chalk line the proper angle needed (center), and then cut the desired shape to fit (bottom).

fitting meant fitting two edges to two or three adjacent stones. At the end, however, we'll have to fill the last open space with a single stone ... and the angles and sides must match all the way around!

In solid construction, this is not a big deal. We would simply mortar in a few smaller pieces at the end (in some inconspicuous spot), and trust the mortar to hold the little bits in place. Laying them out "dry", in sand, is less forgiving. Small, filler pieces can easily work loose in a sand base.

There are two ways to deal with the problem. First, don't leave a narrow space to be filled at the end. If the next-to-the-last rock *almost* fills the space, cut it or discard it. Better to have two medium-sized stones than a big one then a little one to plug the last gap. (Figure 1-6) Secondly, accept that the last rock may require more cutting than all the rest combined. It may, in some cases, be easier to trim the odd projections from stones already laid in order to shape or enlarge the last gap. Again, lay the last stone in place, mark any obstructions for removal, and cut them off. (Photo 1-6)

When the frame is filled, walk back and forth across the areas that will be exposed. If any stones tip, or shift underfoot, reset them with more sand. It might even be necessary to shim up a particularly tippy rock by wedging another thin stone under the unsupported edge.

When all feels solid underfoot, spill another half bucket of sand on the surface and sweep it into the cracks. You don't want the sand to be flush with the surface, but there should be at least enough sand in the cracks to dribble its way into hidden cavities under the stones.

That's it! The four hour hearth. We set up the stove and cleared away the mess. I offered to take home the left-over rocks, but Randy already had the wheeels turning on another stone project for the yard.

Alternatives

Randy's hearth was built to suit a particular situation. If portability isn't on your own list of necessaries, consider some amendments to the Bolger plan:

1) Cut the carpet.   The spring in the carpet pile allows some slight additional flex in the hearth support. In a sand-based hearth, like Randy's, each flexing movement dribbles a bit more sand under the moving rock. So when the living room polka is over, some corners sit a tiny bit higher and the surface will gradually get rougher. In a mortared hearth, any flex in the base risks cracking the mortar joints. Flex, in other words, is better avoided. Rigid is better. Removing the carpet

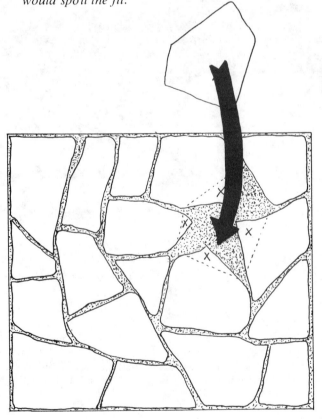

Figure 1-5 and Photo 1-6 The last gap can be the hardest to fill. The solution is to find a stone large enough to cover the gap, set it over the gap, mark around it and cut off the projecting corners which would spoil the fit.

from beneath the hearth would make a durable surface.

2) Change the shape. Use any shape you like — so long as it satisfies safety requirements and the building inspector. Do remember, however, that as the area expands more attention must be given to stopping flex in the bottom of the box and supporting the additional weight.

3) Mortar the stones. Lay out the stones as we did for the sand-based hearth, but omit the sand underneath. Unless the stones are perfectly uniform in thickness, the top surface will be uneven without the sand. The idea is simply to get the pattern fitted. Take the fitted stones out of the box in order, setting them aside so the pattern is not disturbed (turn one stone the wrong way and you'll never get them all in the box again). Wet the stones.

Now, dump a base of stiff mortar in the box and replace the stones, working the stones into the mortar as we worked them into the sand, settling each one down until the top surface of each is level and flush with the rest. Work quickly to finish before the mortar gets too stiff to work. And **do not** stand on the stones — you may tip them or push them too far into the mortar. Work from a plank resting across the top of the box. Once the stones have been set back into the pattern, and the surfaces flush, use the edge of a trowel to work more wet mortar into the cracks between them (see Chapter 10 for more detail).

When the mortar joints have stiffened up a bit, carefully skim off the excess and smooth out the rest with a narrow trowel or a butter knife.

The mortared hearth will be easier to sweep clean, and will hold its surface better than the sand-based variety. I must stress, however, that mortar joints demand a rigid base. Use heavy plywood underneath; and, if possible, put a short cross-beam and a jack post under the floor to take out the "bounce". Ban the living room polka.

23

Chapter 2

Gateposts

The funny thing about a gatepost is the inverse connection between form and function: the fancier the gatepost, the less likely it is to hold a gate. The biggest cattle ranch I ever saw had no gateposts at all. Likewise, for all the years we had livestock here, the gate hung on a half-rotted cedar post. Now that we're down to a pony and a pregnant rabbit, we've replaced the old post with two stone pillars. If the rabbit dies, we may progress to a moat and drawbridge.

All of which is a long-winded way to say that the gatepost is one place where the builder can follow his fancy. Indeed, it doesn't even have to hold a gate. We built the gate hardware into this one mostly to show how it's done. The height, the girth, the connecting fence on either side are all matters more of taste than of technique.

Design

Nevertheless, to make it last, the gatepost design does include two vital features: the base, and the cap. A solid base keeps the post from tipping, and the cap sheds rain so the mansonry won't disintegrate from the top.

The base, for this post, is a "floating" pad. It doesn't really float. It ties the post together into a simple structural unit. When the ground does move, with frost or roots or Doug's truck bouncing up the drive, the whole post rides atop the movements like a raft riding over waves. It may still tip if, for example, the ground freezes unevenly, but it should not crack and break the way a solid rooted building would. A tippy post can be pushed upright again, but a broken post is broken. Period.

Given the choice and the time, I would still prefer to dig down to bedrock, or hardpan, and set the posts on the most solid foundation I could reach. But, some structures are just too small, or too frivolous, to warrant that much trouble. Gateposts are in that class.

A pad will keep it together. Careful placement will improve the odds on tipping. Consider the forces that would upset the post.

First, the traffic. Snowploughs, distracted visitors, delivery trucks, all seem to take direct aim at a gatepost. The design solution is to move them back, away from the traffic. I wanted to hang an existing 14 foot gate in the gap, so that determined the size of the opening. But if you don't intend to mount a gate, take the opportunity to separate the posts more widely than that. The farther from the traffic, the better.

Secondly, heaving frosts don't always act uniformly. This is especially true in a driveway, where the insulating snow is cleared away. Frost penetrates more quickly and more deeply in the cleared part. So you may get heaving frost under the driveway side of the post while the backside, protected by snow, stays put. The result is a leaning post. The solution, again, is to move the post back — away from the area that will be shoveled. Or, now that you have a good excuse, give up snow shoveling entirely.

Third on the list of upsetting forces is subsidence. Uniform subsidence is not a problem. Each pad is 39 inches square, or 10 square feet to support the 4500 pounds of post. 450 pounds per square foot is well on the safe side of the soil's bearing capacity of 3000 pounds per square foot. The danger comes when the pad is allowed to rest on soft spots and hard spots. Hard soil on one side, for example, and soft fill on the other. Dig down past the soft organic soils near the surface, into the subsoil. Scrape the bottom clean and flat, but do not fill any low spots in the excavation. Dig a little deeper rather than fill in the low spots.

Finally, tree roots are prone to get under a post and upset things. You will notice that I have taken the worst of all cases. Not only did I plop these posts under the trees, but one is under a willow tree and the other under a locust — two notorious species for rogue roots. Bad planning, perhaps, but a very good reason to use the floating pad. When the willow begins to throw its weight around, I can sever the responsible root and right the post. The roots may move it, but they will not break it.

The cap is designed to shed water. Water, if it gets inside the post, may freeze and break up the masonry. The solid cap ties the top course together and keeps out water. There are several ways to cap a post. The simplest is to make (or buy) a concrete slab and mortar it on the top like a one-stone final course. The cap for this post, however, would weigh over 400 pounds. In the years since we started stonebuilding, most of our friends have developed bad backs (or so they claim). 400 pounds was more than I wanted to lift, so I poured the cap in place.

The cap, to do its job, should project beyond the face of the stonework, should be slightly domed, and should have a relatively smooth top.

Hinges, if you're really going to hang a gate, take some thought. There are several types available. The cheaper ones screw into a wooden post like a giant lag screw. They have a wide thread and a screw taper. The better ones are made like bolts, and come with nuts and washers. Both kinds work in a stone post, but the nuts and bolts type can be adjusted. The screw type is fixed once the masonry has hardened. In either case, concrete will hold them better than mortar can. We'll make provision for a concrete "plug" to hold the hinges.

An adjustable hinge is desirable for the same reason belts have several notches. Weather, we decided, might move the post around a bit. If the post is on a floating pad, movement is inevitable. The gate, itself, may begin to sag (that's what made me think of the belt). One way or another, many gates reach the point where they no longer swing free, but have to be dragged — digging and scraping — across the drive. That's when it's handy to be able to remove the gate, screw in the top hinge (or reverse the bottom hinge a couple of turns), and level the gate again.

Preparation

Ewan, at that wonderful age between toys and girls, agreed to dig the holes while I built forms for the bases.

I told him to go down at least 12 inches, to chop out all the roots, and not to loosen the soil below the final bottom of the hole. The shape of the hole didn't matter much, since we were using forms. But the bottom did have to be level, undisturbed soil.

One hole bottomed out on hard subsoil at just about the right depth. The other hole, however, was soft, black soil all the way down. That one we dug a little deeper, looking for dirt that wouldn't compress when we walked on it.

Each hole was then filled to within 6 inches of the top with rocks and gravel. We stomped on this fill to settle it and compact it. Then we poured on water to wash the finer grit down into crevices. The idea is to compact the under-pinnings now, rather than leave them to settle later when the gatepost is balanced there.

A gravel base allows drainage from under the pad. Soil would hold the water. And wet soil, when it freezes, expands, lifting a little thing like a gatepost with ease. Gravel is hard, and it doesn't hold the water.

While Ewan was happily soaking and stomping, I assembled two simple forms. Each form was made of four pieces of 1 × 6 lumber, 40 inches long. I overlapped the boards at the corners, nailed them, squared the box, then added a temporary diagonal brace to keep it square. (Photo 2-1)

Photo 2-1   Place the wood frame into the dug-out area and nail on a wooden diagonal brace.

Set the forms on the gravel bases and line them up with one another, and with the driveway. Remember that the pads are wider than the posts will be. If you're allowing a precise gap for a gate, allow for the difference in width.

We aligned these posts by eye. You could do it more precisely by marking out a centre line on the drive and measuring from that; but if your driveway is half as crooked as mine, only a surveyor could tell the difference.

Level the forms. Then dump more gravel around the outside of the forms to hold them in place. When they're secure, remove the diagonal braces. The gravel will keep them square.

Pour the concrete into the forms, and strike off the excess by sawing a long board back and forth across the tops of the forms. Keep the concrete surface damp for a few days while it sets. (see Chapter 14, "Concrete and Mortar")

Post

The trick to raising a post like this is to start square and set the corners plumb at every course. Some masons use a hinged plywood box as a form. They set the stone inside the box, then remove the box to finish the pointing. I prefer to see what I'm doing, however. And — with a little care at the corners — these posts are as straight as anybody's.

Draw the outline of the post on the top of the concrete pad. The corner marks are the first reference points. Lay the first course on the marks and you'll be off to a square start.

The posts shown here are 28 inches square. With something this small, nearly every stone is a cornerstone. (Photo 2-2) And, with something this small, if the cornerstones aren't square it shows. For those who pale at the thought of cutting stone, by all means use the stone as it comes. You will have to work harder at keeping the corners aligned, to make roundish rocks look straight, but the result will still be a gatepost.

Half the fun of stonebuilding, though, is shaping a useless-looking lump into something worthy. Maybe it comes from having a teenager in the house. If I can hammer a square corner onto a stone ... anyway, whether or not you have a deep inner need to pound rocks, here's what to do when you feel the urge.

A cornerstone needs to be reasonably flat on four sides: top, bottom, and two perpendicular faces. If it's a sedimentary stone, sandstone or limestone, chances are that it already has a reasonably flat top and bottom. In that case, the object is to make two faces.

Start with the longest, straightest edge. Lay a yardstick along the top and look for variations from the

Photo 2-2   With a stone project this small, nearly every stone becomes a cornerstone.

line. Using the wide chisel, score the edge where it needs to be straightened. Don't try to take it all off at once, a fault that some strippers have. Just tap back and forth along the scored line, wearing the mark into a rut. Keep the chisel upright, aligned with the face you're trying to make. (Photo 2-3)

With this gentle tapping, close to an existing edge, you may just chip away pieces rather than remove one clean slice. Not to worry. Just chip it away.

After cleaning up the edge, there may still be lumps or protrusions farther down on the face, away from the edge. Without an edge on which to start the chisel, these may be hard to remove. Switch to the narrower chisel, with a blade about 1 inch wide. Work at the base of the bump and turn the chisel at more of an angle — as if you intended to gouge out the bump by its roots instead of slicing it off at the base. Do be cautious, though. If the bump is near an edge, you risk ruining the edge by taking off more than intended. Hammer towards the meatiest part of the rock and away from the fragile edge.

27

Photo 2-3   Shaping a stone requires the handy use of a hammer and chisel. It takes a while, but is worth it.

Having turned a nearly good face into a good one, the next task is to create a second face at right angles to the first one. I use a rather large square here. Lay it on top of the rock, with one side aligned with that first good edge. The other side marks the next cut.

Score a line with the wide chisel. Start gently, until the mark is clear and straight. Rough spots on the top of the rock will twist the chisel a little off the line, so work it back and forth until the line is true. The straighter the line, the better are chances of taking off the excess cleanly, with a single break.

Again, keep the chisel perpendicular to the top of the stone, and parallel with the face you're trying to make. As the hammer blows get harder, the chisel sets up fracture lines deep inside the rock. In fact, you aren't really cutting the stone at all. You're breaking it. And it is breaking out of sight, inside the rock. Keep the line straight, and the chisel straight, and the hidden fractures should come out where you want them.

To improve the odds, turn the rock on edge and score the corner itself. With the cut scored on the top and at the face, pound away. If you've been careful about keeping the chisel straight, the scored line will soon become a crack.

First time stonecutters should not be disheartened with failure. Sometimes rocks break right and sometimes they don't. But you'll soon reach the point where it's quicker to cut another stone than the waste time looking for one that nature made square.

There are two tips that may improve your success rate. First, work on soft dirt, or sand, where the stone will be supported uniformly. On a hard surface, even a tiny pebble under the stone can concentrate the force of the blows on that one small point and break the stone in the wrong place. Secondly, keep the chisels sharp with a regular touch-up on the grinder. A sharp chisel concentrates the force on a smaller area, making each blow more effective.

When you have a supply of cornerstones ready, wet the concrete base and mix up a batch of mortar.

Spread a bed of mortar inside the lines that you marked on the pad. (Photo 2-4) Leave an inch or so between the mortar and the line. As you tap the stones into place, some of the mortar will squish out at least that far.

Photo 2-4   Spread a bid of mortar inside the lines that you marked on the pad.

Set the corners first. Line up each one vertically with the marks on the base. Check both faces of each cornerstone, tapping the stone this way and that until the alignment is perfect and the faces are vertical. I was aiming for a 28 inch square; and so, as a final check, I measured the first course from corner to corner, making sure that it was exactly 28 inches across each face. (Photo 2-5)

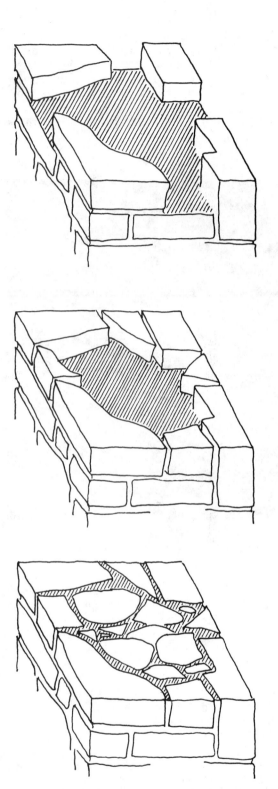

Photo 2-5   Set the cornerstones by lining up each one vertically with the marks you've made on the base.

With the corners in place, fill in the spaces between them. First at the face, and then the middle. The core is no more than rubble and mortar, a place to get rid of the junky stones, and a backing for the four faces of the pillar. Anything goes in the middle, but do proceed in the proper order: first the corners, then fill in the faces, then the middle, bringing the whole course up to a common level before starting with four more corners. (Figure 2-1)

Figure 2-1   Build up the gatepost by first fitting the cornerstones (top), then completing the faces (middle), and finally filling in the center with mortar and waste stones (bottom).

The main reason for leaving the core until last on each course is to leave yourself a maximum of flexibility for fitting the more difficult corners. If there's a big lump of a rubble stone sticking up in the middle, it's bound to be in the way of whatever you want to put on the corner. Fit the corners and the faces, then fit the rubble around the backsides of the good stuff. (Photo 2-6)

Photo 2-6   Fitting in the cornerstone first will make your job a lot easier.

The usual rules of masonry apply: overlap the corners, avoid continuous vertical joints, and lay an occasional large stone on the flat — extending from the face to the opposite side, right across the core. These "bonding units" tie the face to the core and make a more solid structure. The best bonding unit is the concrete cap, which ties the entire top together, but use several bonding units lower in the post, too. If a single stone won't span the entire post, overlap two face stones in the centre. A 6 inch overlap is sufficient. (Figure 2-2)

There is an occasional lost cow who wanders down the road, and the pony looks up from her clover from time to time, wondering, no doubt, if there were another field in the world where she could stay so fat. Strictly speaking, then, these gate posts do serve some function. But you and I and the pony know that the real reason we built them was for looks. So let's consider looks.

Art and masonry come together in the single skill known as the backward step. Take one. Wipe the spatters off the glasses, have a sip of something cold, and look. Look from way back, where you can see the whole thing at once and how it fits the landscape. From that perspective, you can see balance — or the lack of balance — in the work.

Consider three things: the size of the stones, their colours, and their orientation (do they lie flat or stand on edge?). Now look for balance in all those things. Are all the big stones on the bottom, or distributed fairly up the face? Are the bright colours lumped in one spot, or spread around in a softer mottle of piebald wall? Do the corners look weak and pebbly, or square and strong? Would an area of narrow courses be balanced by standing a larger stone on edge now? Does one post look like the other one?

The backward step shows me where I'm going wrong. If I take one often enough I can fix the mistakes as they happen.

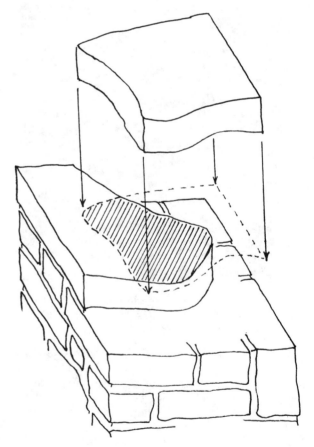

Figure 2-2   Overlapping large stones periodically will help bond the entire unit together.

Hinges

A double-nut, adjustable hinge bolt is too fat to fit in most mortar joints. And mere mortar won't be strong enough to hold the bolt against the pull of a gate and the neighborhood kids who will swing on it. Better to leave a gap in the stones and bed the hinge bolt in a harder, concrete plug.

First, determine a proper height for the bolts. For this gate, I had to set them 11 inches and 43 inches above the base pad. So, when the stonework reached that level, I left a gap between the stones, then filled the core with a concrete mix instead of the usual rubble and mortar.

Work in the concrete with a trowel, forcing out the air pockets and settling the mix into all the crannies behind the face stones ... with one exception. If you work the concrete too far into the gap left for the bolt, the mix will flow out through the gap and down the face of the post. Chop it around, but try to keep the concrete back, an inch or so from the face.

While the concrete plug is setting up a little, take half an hour to prepare the bolt. The object is to leave the back nut bare, so it will be bedded tightly in the concrete. And, at the same time, protect the threaded bolt so it's left free to turn through the fixed nut. That will allow you to screw the bolt in and out of the post to adjust the level of the gate.

First, remove the back nut and washer, and coat the bolt with grease. Replace the back nut and turn it 2 or 3 inches onto the threads. Leave the front nut at the hooked end of the bolt. With the nuts in this position, you will be able to turn the bolt counter-clockwise and back it 2 or 3 inches out of the post. A 2 inch adjustment on the bottom pin would raise a 14 foot gate 10 inches at the far end.

As extra protection, to ensure that the threads aren't jammed by concrete grit, wrap the threaded portions of the bolt with waxed paper or freezer paper, and tape it tightly. Finally, wipe any grease off the nut. We want the nut to be bedded firmly in the concrete. (Photo 2-7)

Now wriggle the bolt down into the concrete. Work it down until the height is right. And work it in until the front nut and washer are right where the face of the mortar should be. The nut and washer won't touch the concrete, which is set back from the face.

When everything is in its place, work the trowel in the concrete that surrounds the bolt. Not much. Just enough to settle the mix tightly around the bolt, and especially around the back nut. Leave no holes.

Later, when the concrete has set more firmly, cover it at the face with fresh mortar. The concrete, made with less lime than the mortar, will cure to a different

Photo 2-7   Preparing the gatepost bolt beforehand with grease will keep any mortar off the surface.

shade of grey. Pointing over the concrete with mortar is a cosmetic touch.

Pointing the ordinary joints is a simple task that can be left until the end of the day (sooner if the weather is hot, dry, or windy). As the rocks are tapped into place, the mortar is squeezed out of the joint at the face. Leave it there, lumpy and rough, until it begins to get stiff. Then force it back into the joint with a narrow trowel, or with a finger (use a glove). Wet your finger and draw it along the joint, packing the mortar and smoothing it. Do not, however, fuss over the joints while the mortar is still fresh. You'll only bring water to the surface and made a mess on the faces of the stones. Leave it alone until it's stiff.

The upper bolt goes in just like the bottom one. Placement, however, is a little more critical. Measure the height from the lower bolt to ensure that the vertical separation is correct. More importantly, use a plumb bob to place the upper bolt precisely above the lower one. You can adjust the bolts in and out, but you can't adjust them sideways. The vertical alignment is crucial. (Photo 2-8)

The cheaper, screw-in pins are set in much the same way. Do not, however, wrap the threads with paper. You can grease them lightly, but concrete around the thread is the only thing holding the pin in place. There is no nut to hold it. Work the pin into the concrete, or "screw" it in, then align it as described above. Don't count on these being adjustable, though. Leave them where they're set.

Photos 2-8 *To place the upper gatepost bolt correctly, first carefully measure the distance from the lower bolt (left), and then drop a plumb bob to ensure vertical alignment (center and right).*

The Cap

If you plan to pour the cap in place, the crucial step is the top course of stone. If it's out of square, or a bump too wide, the form won't fit over the top. On the other hand, if the last course of stone is very much smaller than the form, it leaves a gap which allows wet concrete to leak down the faces of the post.

Measure the top course carefully as you set it in place. You could even slip the form itself over the top, just to check the fit. A small gap is acceptable. We can span that with paper. But any gap wider than an inch is inviting trouble. In that case, you'll have to re-build the form, or re-set the top stones to fit more closely. Re-setting the top course will be easier than re-building the form, as long as you start while the mortar is still wet.

The height of the last course isn't critical, as long as the two posts match. But the forming will be easier if the top edge of stone is more or less level all around. Away from the face, away from the potential gap, the stones can stick up into the concrete. Indeed, it's better if a stone or two does stick up — half in the post and half in the cap. It ties the cap to the post with a solid "key".

The form is less complicated than it looks. Start with a simple box of 1 × 6 lumber. Cut each side 4 inches longer than the face of the post. This was a 28 inch post, so each side of the form was 32 inches long. I overlapped the corners and nailed the box together.

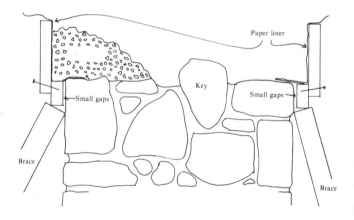

Figure 2-3 *Line the edges of the form with waxed or freezer paper to prevent the wet concrete from dribbling through the gaps between the straight board and the uneven stone.*

32

Overlapping took an inch off the inside dimension. So the inside of the box was now 31 inches. A box, 31 inches square, on a post, 28 inches square, will leave a 1½ inch overhang all around. Form this lip by nailing strips of 1 inch lumber around the inside of the box. (See Figure 2-3)

Slip the form over the top of the post until the inner lip lines up with the top edge of stone. Then prop up each corner with any sturdy lumber that is long enough to reach the ground.

Next, level all four sides of the form. You can raise or lower the form by moving the props in and out at the base.

In theory, we had a 1½ inch overhang in the form and we filled that gap with strips of 1 inch lumber. That should have left a ½ inch gap between wood and stone, all around the post. In theory. In real life, you will almost certainly have to whittle a bit here and there to make it fit. (Photo 2-9)

No matter what you do, there will be a gap of some dimension. Cover the gap with a double layer of butcher's paper, freezer paper, or waxed paper. Cut a piece the same length as the inside of the form. Fold it at the bottom with a double flap, 4 to 5 inches wide. Fit the double flap over the gap and staple the paper to the form. Fit a paper like this on all four sides. Finally, smooth out any creases in paper — these would show on the face of the concrete. (Figure 2-3)

You're relying on two layers of paper to hold the wet concrete above the gaps. So fit it carefully and staple it well, especially in the corners. The wider the gaps, the more careful you have to be.

I have used heavier "tar paper" over very wide gaps. But this proved to be more difficult to remove later. The freezer paper tears away very easily, leaving a neat finish on the concrete. Overall, I find it easier to fit the top course of stone as close to the form as I can, then span the narrow gap with the lighter paper.

Check the level one last time, then pour the concrete into the form. As always, chop it into place with the trowel, being especially attentive along the sides and in the corners, where air pockets would show as holes in the face of the cap. Just don't poke the point of the trowel through the freezer paper.

The appearance of the concrete face can also be improved by tapping the outside of the form with a hammer, all the way around. This vibrates the wet mix, settles out the air pockets, and brings the finer particles to the face.

Though the form looks full, dump one more bucket of concrete in the middle, and work it in with the rest. For this job, we don't want a level top, but a humped one.

The hump changes the way we "screed" the concrete, or strike off the excess. You can't scrape back and forth across the form without removing the hump.

Photos 2-9   Make sure to level all four sides of your concrete form (left), and then line the edges with a double layer of a waxed or coated paper (right). This will give the sides of the cap a smooth finish.

This time, start at the corners and work the excess towards the centre, letting it rise above the forms. The exact shape of the hump is immaterial — so long as it sheds water and looks presentable. Shape the hump by eye, then let it dry for a few hours before you trowel the finish. (Photo 2-10)

Photo 2-10   Shape a slight hump on the concrete cap to aid the runoff of rainwater.

Cover the finished concrete with plastic to keep it damp, and leave the form in place for three days. Then pull out the props and pry the form apart at the corners. The paper tears away neatly and the cap is done.

Alternatives

Lights might be the most sensible addition to this design. We left them out only because the gate was 600 feet from the house, and I balked at digging a ditch that long for the wires. If you're closer to power, or have the energy to dig that far, the lights themselves are a simple addition.

Bury the wire first, in accord with your local power company's rules and permits. Leave a spare 6 to 8 feet of wire at the gatepost end. Then prepare the hole and the pad in the usual way. The gravel base surrounds the wire. The wire (preferably encased in a conduit pipe), is left sticking up through the middle of the base

form. Drive a long stake into the base, beside the wire. Make sure the stake is plumb and centered. Then attach the wire to the stake with twist ties. The purpose is to keep the wire up out of the way and centered while you build the gatepost around it. (Figure 2-4)

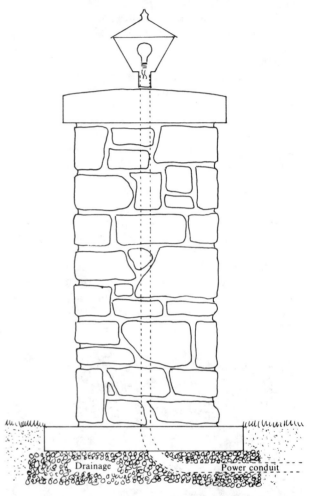

Figure 2-4   To add a decorative touch to your gatepost, you may want to add lights which will require laying a power conduit to supply electricity.

Even better is the commercial, outdoor light, built to fit atop a 3 inch steel post. Use the steel post instead of the stake, and thread the wires through it. The post holds the wire out of the way and vertical while you encase it all in stone. When the masonry is done, you connect the light to the wire and slip it over the projecting end of the steel post.

If the stone supply is a little short, use a big cardboard tube, or even a length of stovepipe in the core. Thread the wires through this central void and lay the stone around it.

The disadvantage of the wide, empty core is the loss of flexibility while fitting stones. If the core takes 10 inches out of a 28 inch post, that leaves a maximum of 9 inches on each side for the stones. A nice, big 10 inch stone would be useless you pared it down to size with the chisel.

There are any number of alternate ways to connect a fence to the gatepost. In part, it depends on the type of fence. Two fences are illustrated here: rustic rail on the left, and a dry stone wall on the right.

The rails are rigid enough to simply abut the masonry, without permanent fastenings. The last post in the fence is close enough to hold them in place.

Some builders are tempted to skip the last fence post and bury the ends of the rails in the masonry gatepost. That's a perfectly reasonable shortcut, except for the fact that the rails rot more quickly than the stones. Replacing rails could be a chore some day.

A better alternative is to bury the heads of two 8 inch carriage bolts in the masonry. Leave about 3 inches sticking out of the face. When the masonry has set, drill holes in a 2 × 4 to march the heights of the two carriage bolts. Slip the 2 × 4 over the bolts, and fasten it there with washers and nuts. Then any type of wooden fence can be attached to the 2 × 4. This is particularly effective with a more formal wooden fence; you can shape or paint the 2 × 4 to match the other posts in the fence. And, when the wood rots, you can undo the nuts and replace the 2 × 4 with ease.

I prefer the dry stone wall. No fastenings are necessary. And, if the frost moves things around a bit, the stones adjust themselves.

This particular bit of wall began with a few nice stones that just happened to be leftover when the gateposts were done. As I cleaned up the mess, the only reasonable place to dispose of the leftover rocks seemed to be the gap beside the post. They looked so nice there that I dug up a few more from the roadside ditch and started a little wall. That looked sort of half-done, so I took Ewan and the truck and went off to gather a whole load of slabs.

The truckload finished the wall nicely. But, with the wall done, there was still one good rock unused. We had hauled it all the way home in the truck. I stood there for a minute or two, holding the lone leftover rock, surveying the scene for a reasonable place to put it. Ewan knew the look on my face. "No, Dad," he sighed. "Throw it away."

He's a good kid, and he had been helping all day in the hopes I would take him for a swim. So I threw the rock into the woods. Not too far, though. I know where it is. I've been thinking that a planter might look nice beside the gatepost.

Chapter 3

Garden Wall

I don't know whether good fences really make good neighbors or not, but a good fence certainly makes a fine backyard.

Tommy and Marion Thompson spend most of the summer in their backyard. It's an important part of their home. But the only way to tell whose backyard was whose backyard was whose was to look for the lawnmower marks. When Tommy and his neighbor mowed on the same day, territorial identity disappeared. One might as well live in a condominium. Marion wanted a wall. Not an unfriendly wall. Just something more substantial than a half inch difference in the grass.

Design

A dry stone wall is a favourite of mine. It's easy to build, yet as solid and satisfying as any more imposing structure. There are a few general principles to follow to make the wall a lasting feature, but the interesting things about this particular wall are the oddities that make it different: the seat, the freestanding end, the nooks for trees, even the chance to mix flat and vertical caps.

The central element of design, however, is the fact that the wall is built "dry" . . . without mortar. The dry wall is easier to build than a mortared wall, and it takes less material. It's broader than a comparable mortared wall, but a mortared wall has to be more solidly based, on a deep foundation. With a dry wall, you can skimp on the foundation, and use that material to make the wall luxuriously fat above the ground.

Any outdoor structure is subject to forces that would move it around: a heaving frost, expanding roots, tiny differences in soil subsidence, even abuse from the neighbor's lawnmover. None of these forces could move a stone wall very far. But, in a rigid, mortared wall, even a little movement is enough to crack the joints. The mortared wall is a man-made monolith. It has to move all together, as a single piece, or it breaks.

A dry stone wall suffers all the same abuse as the cemented wall would. Probably more, since the dry wall has a less substantial foundation and a top that can be pulled apart. But the dry wall can be pushed around without suffering the ill effects. The uncemented joints yield to nature's little pushes and shoves. The stones slide ever so slightly in one direction. Then, when the frost subsides, they slide right back into place. The dry wall can move with the punches. The mortared wall cannot.

Dry wall design must, however, account for the three effects that can destroy the structure: cumulative movements that might eventually overbalance the wall, loose stones dislodged from the face, and loose stones dislodged from the top.

Some of nature's movements are self-correcting. Frost heaves, and then it subsides. Other forces, however, are cumulative. A tree root, for example, grows fatter and fatter. If that root is under the edge of a wall it may eventually topple the wall.

The mortared solution is to cut all the roots and put down a deep foundation. The dry solution is to make the wall harder to topple.

Topple-proofing requires a wide base, a low profile, and some taper. How wide, how low, and how much taper depend on the quality of the stone and the stability of the soil. Large, flat stones make a more stable wall than small, round stones. A wall on an open lawn will be more stable than a wall beside the driveway, or under a fast-growing tree.

A wall with turns and curves is more stable than a straight wall. Each "dog-leg" increases the effective width of the base. Try to balance a playing card on edge. Then fold it once and see if the angle doesn't improve stability.

A dry wall is unlikely to topple en masse. It will tilt so far and then the top, or the face, will fall off. The taper lowers the centre of gravity. More importantly, it leans the face back past the safe side of vertical. So, even if the wall tilts, the face and top have a chance of staying on.

There is no way to state safe limits, or ideal dimensions for something so individual as a dry stone wall. Look at the stones, look at the site, and consider the abuse that the wall must take.

The Thomson wall was a typical compromise. Marion wanted something about 30 inches high. She wanted to keep it as narrow as possible, to limit the amount of lawn taken up by the base. It had to be straight since the property line was. We settled on a 24 inch base, tapering to an 18 inch top. In fact, it's wider than that under the trees, and narrower where lawn space is at a premium. Likewise, the height is only an average. The base follows the contours of the lot. The top, however, was built with the sidewalk superintendents in mind. We "leveled" the top by lining it up with the siding on the house behind, exactly the way the cocktail critics will check the job when it's done.

The capstone course includes three styles: a solid horizontal cap that crosses the top and ties it together, a weaker but more aesthetic top with crannies for soil and plants, and a short section of vertical caps. The solid caps require lots of very large stones. You can use smaller stones for the vertical caps, but you'll need even more of them, and they should be well-matched in size. The planted top can be dislodged more easily than the other two styles, but in as small wall that's hardly critical.

Which brings us to the real virtue of the dry stone wall: it's as easy to repair as it is to build. These walls, if they fall, fall apart from the top. The lower stones are held in place by the weight of the ones above. So the first failure is likely to be one loose stone on the lawn. The solution is to pick it up and put it back on the wall. No hammer and chisel to make it fit again, no mortar to mix, no obvious signs to a patch-up.

The seat illustrates another of the joys of stone wall building: versatility. (Photo 3-1) You can include just about anything that the material will allow, or suggest. Doug and I found that big, flat slice of fieldstone in a meadow several miles away. Weeks passed before he saw the finished wall. The first thing he noticed was the seat. "I wondered what you were going to do with that

Photo 3-1   *Using an extra large stone to add a seat in the wall is an example of the inspirations you can come up with in stone building.*

big one," he said. Doug has been on my stone hunts before. He knows how it works. In truth, the rock had been just a bit too wide for the wall, and it was far too pretty to break. It was the rock that had demanded a seat, not the other way around.

Material

There is only one material in this wall — lots and lots of stone. My guess is about 500 pounds of stone for every linear foot of wall. Tommy and Marion, however, dug most of what we needed out of the flower bed.

The quantity is elastic. You can end the wall where you run out of stone. And add more wall when new stones come to hand.

Quality, however, is more demanding. In a dry wall, bigger and flatter are better. Straight, square edges are a bonus on a big, flat stone, but are not essential. You can make a straight edge for the face with one swing of the sledge. Making a spherical stone flat, however, is considerably more difficult. Making a small stone large is almost impossible. What you're looking for, then are the biggest slabs you can find, and move. Don't worry about the strength or the colour or the shape. Just look for a top surface and a bottom that are in roughly parallel planes.

It's possible to make a wall out of small, round stones, but you'll need mortar to bond it all together, or the patience for continual repairs.

Preparation

You'll need a spade, two stakes, and a string. If the wall doesn't have to be straight, you can dispense with the stakes and the string.

The stakes and string define the line the wall will follow. In this case, it was the lot line. For a precise layout, we would have used two strings, one on either side of the base. But the only real concern was to avoid building on the neighbor's lot. The base, on Tommy's side, could bulge here and there to accomodate bigger stones, or to carry a slightly taller section of wall across a dip in the lawn (the base must be wider in a dip so the profile of the upper wall can remain constant).

If we were building an ordinary, vertical wall, we would raise the string with every course and align the entire face with the string. Here, it's enough to get the base straight. The face of the wall is tapered — it's supposed to lean. So lay out the base with string, and align the rest by eye. Look down the length of the wall. Bulges and misplaced stones will be obvious. If you can't see them, then neither will anyone else. (Photo 3-2)

Use the spade to cut away the sod and level the base. The sod would only decompose, and then com-

Photo 3-2 Align the base of the stone wall with the use of a string.

press. Do be careful, though, to leave the dirt bottom undisturbed. Scrape away loose soil with the spade held flat. Don't dig and loosen more soil than you plan to remove. Loose soil would compact with time, and allow that part of the wall to sink.

Your objective is a shallow trench, flat and hard on the bottom. Soft layers of leaf mould, grass, and roots should be removed.

Fill the trench with a mosaic of the ugliest, most useless-looking stones in the pile. Again, big is better than small. Big, of course, refers to the surface area of the rock, not to its thickness or weight. The idea is to keep the wall from sinking into the soil. Think of a woman walking on snow with stiletto heels. She'd be much better off with snowshoes. What we're looking for here are snowshoe rocks — wide and ugly and flat.

Flat will also be the watchword for the top of this base course. Pick stones of uniform thickness, or sink the thicker ones farther into the soil. Avoid humps, bumps and high spots. They'll make it ten times tougher to fit the next course.

Laying the Wall

The two most useful tools for the job are the yardstick and the hammer. Use the yardstick to find the stones that fit, and the hammer to persuade those stones that almost fit.

The wall is essentially two faces, filled with a rubble core and tied together with an occasional big stone that spans the wall from face to face. The wall will be stronger if you lay every stone "on the flat", showing its edge at the face of the wall, and extending far back into the centre of the wall.

The stones are bound together by the friction between the overlapping surfaces. Overlapping is the cardinal rule. Piling one stone atop another does not make a wall, it makes a pillar. Piling one stone atop *two* others makes a wall.

Pick each stone by these criteria: as high as the stone beside it, shorter or longer than the stone below it. The object is to span the joints in the course below. You can try to reach across a joint with one rock. Or, fall short of the joint and span it with the next rock. Either way, try to avoid lining one crack up with another. Overlap. Stagger the vertical joints.

Match the height, or the thickness of the stones for the simple reason that in the succeeding course you will again be trying to span the joint. If you leave a step between two stones with uneven tops, you would have to carve a matching step in the stone above in order to span the joint.

It is possible to use a small filler stone to change the level between uneven rocks, but if you do this too often the wall gets weaker and begins to look bitty.

When picking a stone, measure the height of the face, and the distance to the next joint. Find a rock the desired thickness, then turn it, or break it, to get a face long enough to overlap the joint below. (Figure 3-1)

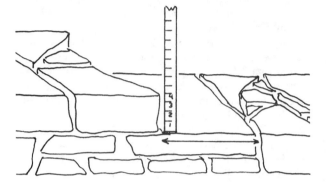

Figure 3-1 When choosing the next stone, measure the height to ensure it is even with the surface, and the width to ensure that it overlaps any joints.

It isn't a meticulous fitting. Close is almost always good enough. In fact, I rarely use the measuring stick anymore, finding that the eye alone is faster and almost as accurate. But you do have to know what to look for: imagine the rock lying flat, match the thickness, don't match the joints.

Fit the more important face first. And you'll have to decide for yourself whether your view or the neighbor's is more important. Then fit the other face. Only by the remotest chance will the two face stones meet along their backsides. If there is a clash between the backs of two face stones, pull out the lesser one, lay it on the ground, and whack it with the hammer. This is not stone cutting. The chisel work is reserved for those stones that have to just right. With these, we're only knocking off their backsides.

Fit the two faces (and if you've done it right, the tops are the same height) then fill the gaps in the middle with any old rock you can grab. Well, not *any* old rock. Stay away from the big ones — those that rise up higher than the faces, making a hump in the middle of the wall. A hump in the middle is almost certain to interfere with the next course up. The centre can be dished — which might mean shimming under the next course — but avoid the convex at all costs. (Photo 3-3)

Shimming is inevitable in a fieldstone wall. Nature doesn't make them perfectly flat. If she did, the rocks would look like concrete blocks, and Tommy's wall would look like Berlin's. Nature's rocks have bumps and wiggles. And when you lay them on the wall, they wobble back and forth a lot. No problem. Put your hands on a wobbly rock. Wobble it forward. Then

Photo 3-3 Fit the face stone of the wall first, filling in the center with waste rock.

wobble it back. When wobbled forward, there's a gap beneath the backside. When wobbled back, there's a gap under the face. Stuff something hard in either gap and the rock won't wobble anymore.

When fitting any dry stone wall, keep a bucket of thin, little rocks at hand. Wobble the rock into the best position, then push these shim stones into the gaps beneath. Given the choice, I would always shim under the backside of a stone. A shim at the face may eventually work free, and then fall out. A shim under the backside, deep inside the wall, has no place to go. It is jammed in position and can't fall out. (Photo 3-4)

Set the face stones, shim them until they're steady, then fill the gaps in the centre. If you have gravel or sand, or even sandy soil, brush some into the cracks atop each course. As the big stones shift, this filler trickles down and fills the gaps under the stones, like so many little shims. The wall gets tighter and tighter with time.

That's the way fitting should go. But, naturally, it's never quite so neat. That's where the hammer comes in. Wall building is much too fast for cutting stone

Photo 3-4 *When fitting a dry stone wall try to make each layer as level as possible by using small stones as shims.*

with the kind of care described in Chapter 14. But a small hammer and a 9 pound sledge are quick persuaders for the reluctant rock.

A rough-edged slab without a decent edge for the face? Knock off a point to square the face. Or whack it in the middle and hope for a straight break — good for two face stones. If you want it to break in a particular line, remember to exploit existing cracks and weaknesses, or use the square edge of the hammer head (see Chapter 13).

A protruding knob that spoils the fit? The little mason's hammer was made for this. Chip off the bump with the chisel end. Again, remember that it's most likely to break on a line parallel to the chisel blade.

Still won't fit? Close, but it's binding somewhere? Smack it on the head with the 2 pound hammer. Something gives somewhere and it drops into place.

Lining up a rock with the face of the wall? You could lift it a little and push it into place, skinning your knuckles raw. Or, you can tap it into place with the hammer.

A hammer is the mason's third hand. So much so

that Tommy teased about the amount of time I spent reaching for it. He painted it white so I could find it more easily. The next day it snowed and the hammer was lost. So was I. When the snow melted, Tommy found the hammer and painted it yellow. Then the dandelions came out. This year he painted it red.

Bonding

By staggering the vertical joints, we've bonded the masonry at the faces. By the same principle, we have to bond the two separate faces together into a solid unit, and we have to strengthen the free end and the narrow sections around the trees.

The best bond within the wall is one large stone that extends from one face to another. (Photo 3-5) The more of these you use, the stronger the wall will be. Look at an older, solid wall of brick. You will see a regular pattern of what appear to be half bricks, mixed in with the rectangles. These are actually ends of bricks, bricks that have been turned across the wall to span the two wythes (or faces) of brick. They are also called "bonding units". Stone works just like that. You

Photo 3-5 *To ensure the strength and stability of the wall you must overlap stones to provide a "bonding" effect.*

don't have to bond in a regular pattern, not for a little wall like this, but do cross the wall with every big stone you can find.

If no stone in the pile is big enough to span the wall, overlap them front to back. That is, set a large stone in one face, then overlap it from the opposite side in the next higher course. A 6 inch overlap in the centre of the wall will suffice.

Bonding is most critical at the freestanding end. Here, we bonded front to back on every other course. Two stones lay along the wall, and one stone crossed at the end. Two more along and one across, like stacking toy logs. We cut the corners reasonably square, and finished with the largest slab we could find, just to hold all the others down.

Mostly because the best and biggest stones were needed on the free end, we started there. We gave it the best bonds we had, then worked our way back towards the trees. (Photo 3-6)

The wall narrows around the trees. That's not a great idea, but it was that or move the property line. We narrowed the wall.

The tree niche is built like the rest of the wall, except that there's only one face. The stones overlap, the joints are staggered, and it's bonded front to back (on either side of the tree). Every stone in the narrow section reaches into the wider wall on either side. The narrow section can't come down unless the wider, more stable, section of wall comes with it. (Photo 3-7)

Capstones

As the wall rises closer to the final height, leave room at the top for the capstone course. This top course is special. Aesthetically, these stones are the most visible ones in the wall, and a well-matched line along the top does more than anything else to give the wall a neat, orderly appearance.

Functionally, the capstones are even more important. A hefty capstone holds the smaller rocks below it in place. The wall can't come apart unless the capstone moves, and the bigger the capstone the harder it is to move. More than its sheer, dumb weight, the capstone acts as a bonding unit, tying the front face to the back.

In the best-built wall, every capstone would cross the whole width of the wall. That requires more large stones than we had for Tommy's wall, however. We had an 18 inch top to span. And a large part of the biggies were busy bonding the lakeshore end and the tree niches. By the time we got to the top of the wall, the best we could manage was full-sized cap every 4 or 5 feet.

When the stone supply doesn't provide enough of the best, concentrate the full caps where the wall is

Photo 3-6

most abused: open stretches where it's likely to be climbed, exposed ends where passersby might prod, and wherever subsidence might be a problem. The Thomson's wall will probably lead a quiet life, so the scarcity of full width caps didn't worry us. We spaced the best caps evenly along the wall, and took comfort in the fact that this 30 inch wall would be too high to sit on, and too low to lean on.

Between the full width caps, we pieced in the largest rocks we could, taking care to turn them across the wall wherever possible. Those that won't span the top should at least be turned so that they extend their length into the centre. The longest edge, in other words, runs at right angles to the wall. This makes it very much more difficult to dislodge.

The patchwork top provides the cubby holes for planting. Marion had a moss that we tucked in here, but I would also like to try rockcress, and a perennial creeping phlox. Wall gardeners should keep in mind that the wall drains quickly (if it didn't, frost would be a problem). Select plants that thrive in dry soil, or resign yourself to watering the planted wall as regularly as you would a hanging basket.

Photo 3-7   Trees need not alter your planned wall line. Tree and stone can co-exist if you allow for "tree niches" along the wall when necessary.

Vertical caps — flat stones set "on edge" — make a fine topping for a wall. It does, however, eat up good stones even faster than a horizontal cap. Matching stones is an aesthetic consideration, but not a functional one. Indeed, some builders prefer a rough top, with pointy corners sticking up in shark-tooth style. It would discourage climbers and sitters I suppose.

Functionally, the important dimension is the same for vertical caps as it was for the horizontal ones: the width of the top of the wall. The object is to span the top with every stone. Selection, then, involves matching the long edges of the slabs (each stone will be stood up on its longest edge).

Behind the seat, the wall narrowed to a 12 inch top. So, as I worked towards the seat, I began to save those stones that had a 12 inch edge and two flat sides. I stacked them one atop the others. By the time I was ready to cap the wall, the stacks were 3 feet high. I had enough matched stone to build 3 feet of vertical caps along the 12 inch wall.

Selection is the hard part. The building is easy. Start with one solid end. Stand a stone on its edge and push it hard against the abutment. Hold it there with one hand while you stand another stone in place. It's exactly like filling a bookshelf — don't let go or they'll all fall down.

When you get to the end, or when you feel like a rest, hold up the row with a big "bookend" stone.

For the sharktooth style, that's all you need to know. Compulsive tidiers, and those with neat bookshelves, will want to line them up a little straighter. My excuse was wanting a more comfortable back for the seat. Here, shimming is not the complete solution that it was lower in the wall. Start with a re-ordering of the stones. If one stands too high and another too low, swap places. Shuffle the deck until the lineup is as straight as you can make it. At this stage, you should still be able to pull one stone out of the pack fairly easily. If the fit is already too tight to shuffle the order, move the "bookend" out a little.

Now try cutting off anything that protrudes from the desired line. Mark the stone first, then pull it out of

43

the pack for cutting.

As a last resort, for those that just won't sit right, shim them up into line. Save the shims until last for several reasons: First, no matter which end you shim, the shim will be at a face, exposed and vulnerable. Secondly, there is less total weight on the shim than there would be lower in the wall; the less weight on the shim the more likely it is to come out. Third, the purpose of the cap is to bond the top of the wall; if the cap is balanced on a shim, it's not as effective as a bond.

When the fit is satisfactory, you're ready to tighten it up. Start at the solid end, pulling each rock tight against its neighbor, taking up the slack one stone at a time. Do this in sections, a few feet at a time. A longer row of vertical caps would be too hard to move and too hard to hold in place. When you've pulled a section together, tap the last one gently with the sledge. Repeated, gentle raps are better than one big rock-shattering swing. Now, replace the "bookend".

With a smaller hammer, go over the exposed edges of the capstones, tapping them into alignment. Most of the aligning has already been done, but this final hammering may still settle them into a closer fit with adjacent stones, creating more slack in the line. And the last step, naturally, is to take up any slack that might have been created. Shove the big "bookend" hard against the end of the line. If you have any doubts that the mass of this last stone is sufficient to hold the row in tight configuration, replace it or build around it with more big caps. (Photo 3-8)

The test is to try to remove one of the vertical caps. If you can lift any one of them out, tighten up the row some more.

Alternatives

Given the time and the stone, the wall could be capped vertically for its entire length. It would take one big, solid horizontal cap at either end to hold the row in place. There would be a payoff in longevity. Every wall needs minor maintenance, replacing the fallen stones in spring, but well-capped walls last longest.

It is also possible to use concrete or mortar in the cap course. Mortar between vertical caps guarantees the tightest possible fit. Mortar beneath horizontal caps improves the bonding across the wall. A low mound of concrete along the top of a wall can act as a complete cap course in itself. I don't have to tell you what it would look like, but, functionally, the concrete cap does the same job as a stone cap.

Any mortar or concrete on the wall will crack as the

Photo 3-8   Fitting the capstone will ensure stability of the wall.

wall moves. Unless, of course, it is stacked on the wall in movable units, like concrete stones. This is one possible solution for those who don't have enough material for a proper capstone course. Make cap "stones" in a concrete form and top the wall in formal style. (Figure 3-2)

The width of the dry stone wall allows any number of extras. I like the combination of greenery and stone, and would probably include a larger planter in another wall like this. You could build a bird bath into the top. Or include posts and a gate (see Chapter 2).

Since the wall is dry, additions can be introduced at any time. Pull out a section of wall and stick in whatever the backyard needs.

Like the Thomsons. When we first built the wall, Marion wanted to end it under the pine. When we got to the pine, there was a small pile of leftover stone. Tommy wasn't sure what he wanted to do with it, so we left it there in the middle of the lawn.

I didn't get back to visit for a month. By then, Tommy had built the leftovers into a brand new section of wall. Then he had unearthed another supply from the old flower bed, and Marion was down at the lake, pulling out more stones for the wall. The little wall was a big wall, half way to the house and still growing. Stone does that to people.

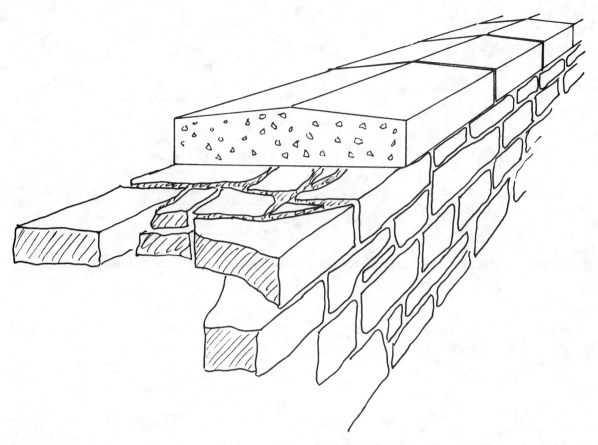

Figure 3-2   You can give your wall a more finished look by pouring a concrete capstone. However, it is a lot more work to pour concrete for the entire length of the wall. You'll have to judge for yourself whether it is worth the extra effort.

Chapter 4

Barbeque

When Dan suggested we build a barbeque, I winced. The last time he told us they were having a few friends over, it evolved into eighty guests and a whole roast lamb.

The challenge was to design a simple backyard cooker that would do for quick, family burgers, and yet expand on occasion for big parties. Moreover, we wanted to include all the versatility that comes with a wheel-away store model barbeque: adjustable cooking height for temperature control, and removable grill for cleaning. This one would do everything theirs could do except rust.

And, in the same way that a casual cookout becomes a whole beast feast, we decided to build with the odd-shaped lumps of granite that cobble the Maruska lawn ... just to make it interesting.

Design

The core of any cooker is the firebox. It's also the logical place to start planning. Plan the firebox around the grill size, and — in this case — around the spit. Dan had a grill 23½ × 17½. So we planned adjustable mountings 22½ inches apart. The grill would slide through this opening with half an inch rest on either side. (Figure 4-1)

The grill, on its own, wouldn't need a deep firebox. A shallow bed of coals and a few inches of adjustable cooking height would suffice for the burgers. But the firebox had to be deep enough for a log fire under the spit. So we would need a removable pan to raise the coals under the grill, and adjustments to lower the grill onto the coals.

The 51 × 24 inch area of the main firebox was an estimate of how much room we needed for a lamb or a small pig. The problem was what to do with all that extra firebox between major feasts. The solution was to cover both ends with removable counter tops. This provides working surface right where it's needed most: at the chef's elbow, and higher than most of the neighborhood dogs.

Some builders will find this firebox more than they really need. Few, however, would spurn the generous working surface. At the end of this chapter, we'll look at a modification of this design that combines a burger-sized firebox with these gourmet counters.

The area of the firebox is determined by the size of the meal, but the height is determined by the size of the chef. Pick a comfortable working height, and measure everything else from that level. A kitchen range is about 36 inches high. We started with that, then decided on an appropriate firebox depth.

Using firebrick, height is usually in multiples of 5 inches. (Each brick is 4½ inches high, plus up to ½ inch for the mortar joint). We chose a 20 inch depth for a lamb on a log fire; 15 inches would have been too hot at times, 25 inches too cool. A 36 inch working height, minus a 19 or 20 inch firebox, meant that we would need a 16 inch base to set the firebox on.

Dan had a pile of used concrete blocks around and the cheapest solution, for him, was to use these blocks in the base. Two block courses, without mortar, would raise the base exactly 16 inches. And the top of the blocks would make a flat, level base for the firebricks.

Don't overlook the importance of this vertical planning. The area of the firebox can be big enough for an ox or small enough for a weiner roast. But if the height

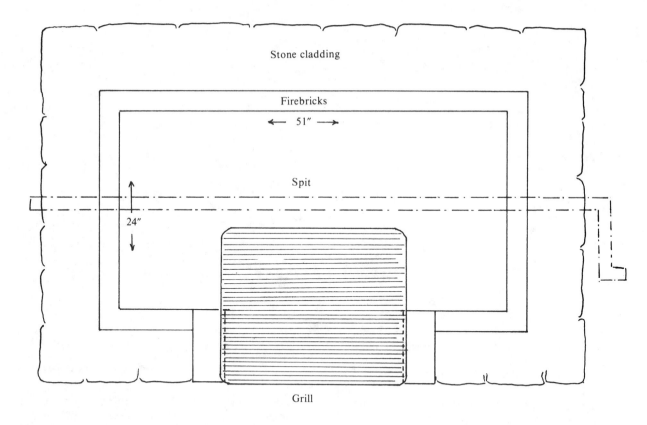

Figure 4-1   Plan the firebox around the grill size. In this case the grill was 23½ inches by 17½ inches.

is wrong, the chef's bent back will regret it.

The height that is not critical is the height of the back. The raised back wall serves primarily as a wind screen, secondly as a heat reflector, and finally as a place to hang the spit. Once the spit hooks are mounted, you can raise the back wall to whatever height looks right.

The labour-saving feature, the one that makes this a two day job instead of a four day job, is the floating pad. Dan's soil is deep, and so is the frostline. A proper foundation, at the proper depth, would have placed more masonry underground than above. So we designed it to sit on a solid concrete slab. The frost will still heave beneath it, but the structure will ride out the movements on its concrete raft and emerge, in the spring, unscathed.

In this type of structure, the potential for frost damage comes from the other direction — from above. Without a roof, rain can fill the firebox and saturate the ashes and the base. The pad is solid concrete, so the water can't drain straight through. When the trapped water freezes, the masonry cracks.

The design must include some means for the water to escape. Our choice was to lay the block base without mortar, then drain from the top of the pad with a length of hose on every side. Water runs through the blocks and out the hoses. Had we chosen to mortar the blocks, we would have had to drain from above the blocks.

Materials

Concrete blocks, for the base
Firebricks, for the firebox
Gravel, for drainage under the pad and for concrete
Sand and cement
Stone
Spit hooks
Short lengths of hose
Brick ties

We used concrete blocks because they were there, 28 blocks in all. Discarded bricks would work as well. Or flat stones — provided you capped them with concrete to make a perfectly flat base for the firebox.

Firebricks are 9 × 4½ × 2½ inches. (You can buy a thinner, one inch brick, but I would not recommend it

48

in a box this size.) This big box took 77 bricks altogether. Half bricks and odd dimensions are easily cut, but other three or four extra bricks to replace ones that break the wrong way.

Some builders substitute cheaper materials for firebrick. Some grades of clay brick (often sold for chimney construction) will last in a firebox. Even concrete blocks are used. But none of these materials will stand up to the same heat as a firebrick. Most stone will take very little heat, and an unlined barbeque will have a very short life.

Any sort of gravel will do for drainage. Concrete gravel should be in the proper proportion (see Chapter 14), but a structure like this is not going to test the limits of the concrete. You won't need much — less than a cubic yard for both pad and drainage.

Sand and cement, too, are needed in small quantities. If you're mixing your own concrete, you'll need about two bags of portland cement for the pad; and save any leftover portland for the firebox. For the stone masonry, quantities of sand and cement will depend on how closely you can fit the stones. If you don't have a sand pile in the yard already, this may be the time for bagged mortar mix.

There's much less stone in this project than there seems. It looks massive, but that's a 7 inch veneer around an almost hollow core. We picked up all we needed from Dan's lawn. If you have to haul it, select fairly uniform shapes and sizes, and save the flattest rocks for the top ledge.

Dan had the spit hooks made at a local welding shop. With a torch, a vice, and a hacksaw, you could make them yourself from any piece of flat steel. Start with the spit itself, and make the hooks to fit the spit. (Figure 4-2)

I put "hose" on the list because most garden sheds seem to have a coil of leaky hose somewhere. The object, though, is simply to leave small diameter holes through the wall. Any scrap of pipe or tubing can be mortared into the wall for this purpose. Cut pieces long enough so that neither end will be plugged with mortar.

A dozen brick ties will be plenty. They're not really essential, especially if your stones are flat (or well-fitted). But, for a few pennies each, brick ties make cheap insurance.

Photo 4-1   Begin your barbeque project by making a level gravel base about 12 inches deep.

Preparation

Start with a hole, about 12 inches deep, and slightly wider than the pad will be. As you approach the bottom, use the shovel on the flat — scraping rather than digging out the dirt. The object is to leave the bottom of the hole flat and undisturbed, with no loose dirt. And, the bottom of the hole could be humped a little in the centre, cut a little deeper at the sides. Unless the soil is porous, a dip in the middle might allow water to collect under the pad. We want to get rid of that water.

Fill the bottom 6 inches of the hole with coarse gravel. Tamp it down as you fill.

Next, make a simple box form for the pad. Four 1 × 6 boards will do the job. Cut each board 1 inch longer than the side of the pad. The excess allows an overlap at each corner. Nail the corners, then square the box and nail a diagonal brace across the top to keep it square. (Photo 4-1)

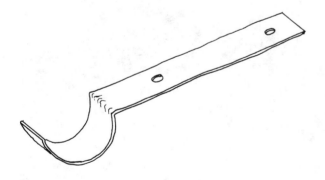

Figure 4-2   Spit hooks can be fashioned out of any piece of flat steel.

Set the form on the gravel and level it. Place the level atop each side in turn. Tap the high corner down into the gravel until the level says it's horizontal. Or, raise a low corner and push more gravel beneath it. When the entire form is level, shovel dirt and gravel all around the outside to hold it in place. Now, remove the diagonal brace (carefully ... you don't want to pull it out of place after going to all that trouble to level it).

Hose down the fill and the form, then get the concrete ready (see Chapter 14). Four parts gravel to three parts sand and one part portland cement will be strong enough for this pad.

Pour the concrete into the form and knead it into place with a shovel. Work out the air pockets and settle the mix into the corners.

When the form is full, strike off the excess concrete with a board. Slide the board back and forth across the top of the form, pushing a wave of wet concrete ahead of the board. This excess fills any low spots, and finally spills over the end of the form. (Photo 4-2) Repeat the process in several directions, until the surface looks flat and wet. When the visible water has soaked away, cover the surface with a sheet of plastic, to keep the top from drying out too soon. Then leave it alone for at least a day.

When you do return, you can remove the form if you wish, but the lumber won't be good for much else. You might just as well leave it in the ground.

Raised Base

This is little more than a filler. It's faster than stone, open for drainage, and provides a flat base on which to build the firebox.

Begin by marking the outside dimensions of the firebox on top of the concrete pad. The firebox has to fit on the outer rim of the raised base, so the base has exactly the same dimensions as the firebox. The pad, of course, is larger than the base; there's a 7 inch lip all around, to support the stone veneer.

Now, lay out the blocks inside the marked perimeter. We set the first course properly — with the holes up. Then we filled the cores with gravel. The gravel provides some ballast, to limit the damage that lead-footed masons might cause. And, it assists the drainage. The rain water that comes through the top will bring ash and other debris with it. The gravel helps keep the spaces open and the drainage flowing.

The second course of blocks were laid on their sides,

Photo 4-2   Level the concrete in the base by sliding a flat board across the top of the form, pushing a wave of concrete in front of it.

presenting a solid top. This makes a better base for the firebricks, and will make it easier to shovel out the ashes. (Photo 4-3)

Photo 4-3   A flat second layer of blocks will provide a good base for the firebrick layer.

The usual rule of masonry applies, even in this kiddy-block core: overlap, staggering the vertical joints.

Because the over-riding dimension is the firebox (based on a 9 inch brick), the blocks (16 × 8) are unlikely to come out even. No problem. Put full-sized blocks at the corners, and fill the last gap with a cut block, trimmed to size. Cut the blocks with the wide 3½ inch chisel. First, mark the line, keeping the cut opposite one of the core holes if possible. Score the line with the chisel, then flip the block over and score the other side. Three of four blows on each side should be enough. Don't worry if the fit is a little sloppy. A little space between blocks is better than a tight fit.

When the base is assembled, lay out the drains with one end of the hose at the edge of the pad, and the other end tucked into the bottom block course. One drain goes at each side of the structure, wherever you can fit a hose end into the blocks. If you're building on a slope, it is especially important to keep the downhill drain open and free of mortar spills.

Now, start the stone exterior. Wet the concrete lip,

butter it with a thick bed of mortar, and settle in the stones. Raise the stonework almost to the top of the base. Let the excess mortar squish out behind the stones and up against the blocks. Pack more mortar behind the stones, filling the gap between stones and blocks.

When these first stone courses have set, they will lock the base together and stabilize it. Even without any mortar of its own, the block base will provide an immovable foundation for the firebox.

More on how to lay these stones later. Once the base is locked into place, we leave the stonework to assemble the firebox.

Firebox

The thing to remember about firebricks is that they're dryer than a temperance toast. They can suck the water out of mortar like a mouthful of alum. You must soak the bricks before laying them. Period.

Even with wet bricks, you have to work quickly. If you're unaccustomed to this sort of masonry, stack the bricks up first without mortar — just to make sure that everything fits. (Photo 4-4) Remember to leave spaces for the mortar, and lay the bricks in the familiar overlapping pattern, bridging all the vertical joints. The critical dimension is the opening at the front. This space must fit the grill. So start at the opening, and stack up the entire box.

Photo 4-4   If you're unsure of the brick design, first lay out the brick without mortar.

If any bricks have to be cut, cut them now. Use the wide chisel to score a straight line on the face of the brick. Now, one sharp rap will break it. The sharper the chisel, the cleaner the break. When the half bricks are cut, and you're familiar with the pattern, dismantle the box and put the bricks to soak while you mix the mortar.

High lime mortars, like those in masonry cement, may degrade at high temperatures. There are commercial, pre-mixed mortars made for high temperatures, but these may not weather well in an open-air barbeque. Here, portland cement is the most durable compromise. Mix three parts sand with one part portland cement, and make the mix a little creamier than usual.

Set the bottom of the opening first, mostly to ensure that it is centered and is exactly the planned width. Level it, and align it along the front.

Now, spread a narrow mortar bed along the rim of the base. Set the first brick in place and tap it down until it's level and plumb. Butter the second brick on its end before setting it. This puts mortar in the vertical joint between the two bricks. Slap some mortar onto the end of the brick, then trim it off with the trowel. The trimming is to keep a uniform dollop of mortar on each brick. If the quantity of mortar is constant, the size of the joints will be constant.

Tap the second brick into the mortar bed, and tap it into its butt joint with the first brick. Level and plumb.

Finish the first course, and start a moortar bed for the second, along the tops of the first row of bricks. Don't lay out the mortar bed too far ahead of the work. It may be stiff before you get to it. Remember that the second course will be a little different from the first. Where you started with a full brick last time, you'll have to start with a half brick now to keep the vertical joints staggered.

Set some brick ties in the mortar bed atop the second and third courses. Space them around the box, with more in the long wall than in the short walls. Hide one end of a tie between the bricks and leave the long end flapping in the space where the stones will go. Set two more ties on either side of the opening. We'll use these last ones to tie the firebox wall to the bricks around the opening.

Lay up the bricks around the opening with special attention to the leveling and plumbing. If these aren't properly spaced, it will be difficult to fit the grill. As you reach the ties on either side, incorporate them into the mortar beds.

Don't be concerned if the bricks around the opening rise higher than the sides of the box. We'll make up the difference later, with a stone cap atop the firebox walls.

When you reach the lowest grill position, make sure the two bricks on opposite sides of the opening are level with one another. (Photo 4-5) Then spread a narrower than normal mortar bed atop the bricks, so that even after you squish the next brick down, the mortar won't come all the way out to the edge. Squish the next brick down, level it, then clean out the mortar joint so that the grill will slide through easily. You can tamp the mortar back in the joint with the edge of a narrow board; then wet the board and slide it back and forth in the joint to smooth the mortar in the channel. When all the bricks are set, and before the mortar hardens irrevocably, you can put the actual grill in place and make sure there's room for it to slide in all positions.

When the firebox is done, clean the inside by slicing off the excess mortar with the side of the trowel. Now, if you like, you can run a wet finger along these joints to smooth them.

Photo 4-5   Make sure the bricks on each side of the grill are level with one another.

Stone Exterior

We laid up this shell with rough, granite lumps. It's not an essential part of the design. You can use any sort of stone. But the boulders gave us a chance to talk about some of the problems posed by the material.

All of the usual rules apply: start at the corners then fill in the face, overlap joints, and tie the face to the rest of the wall. But now those rules meet new conditions. The corners are square and the rocks are round. What gives? How do we overlap joints when the top of each course is a roller coaster? What's to tie when the face is all there is to the wall? The entire wall is only one rock thick.

Let's start at the corner. Look for a boulder with at least one flat side. Roll them around. If you can find a flat side, you're halfway home. If you can't find one, you'll have to make a flat side. It's probably not true, but let's pretend they're all round.

Look for the boulder's grain. You may have to wet it to see the faint, parallel lines of colour that often mark the grain. We want to split the rock along those lines.

The fast way to split it is to hit it with the sledge hammer. Better still, hit it with the front edge of the hammer's face (see Chapter 13). This part of the hammer acts like a big, fat chisel. Line up the front edge of the hammer head with the grain of the rock, and swing in the same direction that the grain runs through the rock. Hit it three or four times in the same spot. If it doesn't budge, you may have to use a smaller chisel.

Smaller? Doesn't the tough one call for the big guns? No. The tougher the rock, the smaller the chisel. The smaller blade concentrates the hammer blows on a smaller area. The force is more intense, more sharply directed. Think of a nail ... it's easier to drive in the little end than the big end.

Place the 1 inch chisel parallel to the grain. That is, the blade should be parallel to the grain, and the shaft should be parallel to the plane that runs through the rock. We're going to use the chisel like a wedge, to drive the two halves of the rock apart. Start slowly. The first few blows are just to chip a little niche in the rock. When it's started. Pound away ... as hard as you can!

When it splits, and it will, you should have two rocks with flat sides. You are twice as blessed as the guy who found one flat side five paragraphs ago. One flat side makes a dandy face stone, but it's a corner we're after.

This time, forget the grain. The object now is to cut a new plane, perpendicular to the first flat side. In other words, we're going to quarter the original round boulder.

Score the cut with the wide chisel. The wider the chisel, the straighter the line. Keep the shaft of the chisel perpendicular to the flat face. Work back and forth along the line. As the line becomes a groove, hit harder.

If the wide chisel doesn't work, switch to the narrow one. The narrow chisel has more cutting power, but it increases the risk of the break running off the line. Keep moving it along the line, and keep the shaft perpendicular.

When all else fails, you can still smack the half boulder with the sledge. Again, try to line up the front edge of the hammer head with the intended "cut". The sledge is hardly a precision instrument, but you can affect the odds by lining it up. If the rock doesn't break along the line, it will at least break somewhere. One way or another, we're going to end up with a quarter of a boulder (perhaps an eighth!). The broken planes, however, may not be at right angles to one another. The right angle maker is the 1 inch chisel. Chip away at the edge, flaking off the excess in small bites.

Now the stone has two perpendicular planes. But it's not quite a cornerstone yet. It lacks a top and a bottom. Take off the pointy ends with the chisel. It doesn't have to be much — just enough to balance it. (Photo 4-6)

Photo 4-6   Chisel off the rough edges to make the cornerstone as neat as possible.

That's the hard way to make a cornerstone. But, when working with boulders, there are only three alternatives to the hard way: 1) look for naturally flat planes and break perpendicular to these; 2) break up the boulders with the sledge and hope for the best; or 3) turn the rounded side of the rock out and live with a lumpy face.

Dan and I built the barbeque largely with alternatives one and two. Even the most hopeless-looking rocks offer a few flat planes. And a fast sledge produces some good breaks along with the bad. The bad breaks get built in somewhere else. And we cut a few the hard way.

So much for corners. What about the rest of the wall? How do we span joints in a jigsaw course?

The easiest way is to stick to fairly uniform sizes. The valley between two round shapes makes a "V" shaped nest for the next rounder. If it's in the "V", it bridges the joint. And, if it's about the same size as the two beneath, it will extend halfway across their tops to meet the stone in the neighboring "V".

But boulders don't come perfectly round, anymore than sandstone comes perfectly square. The natural assortment is a hodge-podge of odd shapes. In practice, it's a matter of turning the stone to find the face, then matching that face to a space in the wall.

Finally, consider the depth of the wall. The best stone is one that reaches across the full depth — in this case, 7 inches. You can chip off the backside of a too-fat rock. But a too-thin rock has to be balanced at the face and filled in behind. When you turn the rock to find a face, don't look for the broadest face necessarily, but the face with a solid, 7 inch backside. Its staying power is directly related to how far back in the wall it reaches.

Fit a stone first, then lift it out and slap in some mortar. Tap the stone into the mortar bed, aligning the face with the wall. Here, where the wall was backed by the firebox, we knew that the face should always be 7 inches out from the firebox. So, instead of the usual string, we aligned the face with a ruler. Just wiggle the stone around until the face is plumb, and 7 inches away from the firebox.

Fill in around the back of the stone with smaller rocks, making a solid base for the next higher course. But don't try to pack mortar into the gap between stone and firebox. If some mortar squishes through the stones and touches firebrick here and there, fine. Just don't fill the space completely. Fill as much of the 7 inches as you can, and still leave a little air gap at the back for heat expansion.

When the rock face meets a brick tie, sticking out of the firebox, bend the brick tie into the mortar bed, put a little more mortar atop it, and then a stone.

The narrow gap between brick and stone must be closed at the top. Otherwise, water will enter the gap, freeze, and expand. That would be worse than the heat expansion. Cover the gap with a cap course.

Use flat stones for the cap, or bricks, even leftover firebricks. A flat surface isn't essential, but the chef will appreciate a place to set his glass. Try to span the top of the wall with a course of single width stones. This is not the place to piece in behind the face with small fillers. Ideally, each stone would be wide enough to reach from the face to the top of the firebrick, covering the gap. But we don't want to cover the top of the firebrick completely, since we need a little ledge to hold the wooden counter tops. For our 7 inch wall, we needed capstones between 8 and 9 inches wide. Dan had some flat limestone around; but, if we hadn't had the limestone, I would have been tempted to substitute some of those 9 inch firebricks.

The back wall must also cap the firebricks. When the stonework reaches the top edge of the firebox, switch from a maximum 7 inch depth to an 8 or 9 inch stone. From here on up, the wall overlaps the top fo the bricks.

Set the spit hooks level with one another, and centered along the firebox. We set the first hooks, then put the actual spit in position to check the level and alignment.

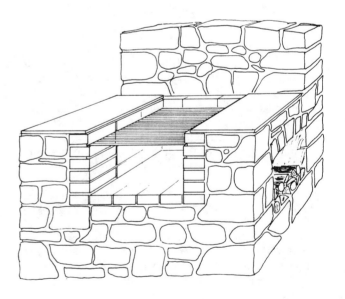

Figure 4-3   An added firebox for additional cooking capacity, or for use as a working oven, would be a convenient feature.

The hooks must be cantilevered to hold a loaded spit. In other words, put a lot of weight on the back end of the hook. Rely on gravity, rather than mortar, to hold it down. When you have at least one heavy stone above the highest spit hook, you can top the back wall anywhere or anyway you like.

Point the face joints, then point the surface joints in the capstones "flush" (see Chapter 8 or 10). Cover the barbeque with plastic, and refrain from lighting a fire for at least a week.

Alternatives

A reasonable modification, for many, would be a smaller firebox. If your cookouts are small, you could eliminate the long box, and the spit hooks. The counter tops, however, would be useful in any barbeque, as would the wind screen back.

To scale down the size of the firebox, plan it around a specific grill size, then match the block base to the size of the firebox. The areas under the countertop ends could be solid masonry, which is easier than the fussy, veneer fitting. Or, use the space for storage bins. (Figure 4-3)

One could easily add to the grill space in Dan's plan by mounting a second grill behind the first. This could stand on legs from the firebox floor, or rest on removable rails, across the top to the firebox. I would be tempted to mount a second grill higher than the first one, and use it as a warming shelf, or back burner.

Chapter 5

Patio

Every house has its ugly side. This was ours. The power company said that the electric meter had to be there. The water tap collects hose tangles the way the closet collects hangers. Extension cords hang around the outside plug. Even the well was ugly. We tried to bury it once, but it took revenge by breaking in the middle of winter.

The best we could do was to concentrate all these ugly essentials in one spot and do what we could to hide them. We couldn't put them behind the house — the meter man wouldn't walk that far. And the hose and extensions are in constant use. Worse, all that traffic and dragging hoses wore away the grass and anything else we tried to grow here.

The answer was the utility patio. Nothing fancy — just a durable stone surface and some strategic planting beds. The mini-cellar that we had to make for the well would make an equally servicable root storage bin, or underground compost maker. More on those alternatives at the end of the chapter. For now, let's assume that there is an actual well to live with.

Design

It was the week before Christmas, and all through the house, not one tap was dripping ... and yet, the pump kept turning on. There had to be a leak somewhere. After dismantling every pipe, from the taps to the hole in the wall, and reassembling it twice, I had still not found the leak. At the absolute nadir of a cursed plumbing funk, I opened a bottle of wine, hoping to find the plumbing muse there.

At the top of the cork, who should arrive but Donald, blacksmith and occasional plumber. Somewhere during the second glass, the pump kicked on again. Don cocked his head to one side, listened, and oracled: "Sounds like a crack in the intake pipe; probably in the well, at the water level."

I am usually dubious of these folkloric revelations. Don also predicts the weather a season ahead. But when he asked me if I bought my pipe at a certain cheap outlet, I figured he might be onto something. The store guess was considerably more accurate than the weather predictions.

Unfortunately, the only way to check this particular bit of free advice was to hack through three feet of frozen soil to the well, which had been hastily buried before the fall freeze. We did it, though, and learned three things in the process: (1) Don has an excellent ear for plumbing, (2) "White Christmas" is a much overrated piece of misguided sentiment, and (3) neverne-vernever bury the well again.

We would insulate the well and intake pipe with something more malleable than winter soil; and we would cover it with a door; and we would not use cheap pipe again.

The well hole is a double-walled crib. The inner liner is local cedar, chosen for its resistance to rot. I was tempted to skip the wood and try an unlined stone crib instead. The liner, however, has its advantages: (1) it's easier to make a straight top edge with wood, and a straight top would be essential for a snug-fitting door, (2) it's easier to attach hinges to a wooden crib, and (3) a wooden liner would keep out soil that might wash through the cracks of a dry stone crib. I didn't want to mortar a stone crib in case the well ever needed major work. For a utility patio, we thought that adaptable might be better than permanent.

Then why add stone cribbing outside the wooden liner? Those who didn't freeze their hands to the pipe that Christmas might call it traumatic overbuilding, but I do have some reasons that have nothing to do with winter. First, dirt has an almost insatiable desire to fill up holes. It can get downright pushy about it. I had doubts about the ability of a simple cedar box to hold back the lawn over the long term. Even a little bit of bowing in the timbers would ruin the fit of the door. The stone would hold back the pressure without yielding.

Secondly, when the liner does eventually rot away, the stone cribbing will protect the integrity of the hole until we replace the liner.

Finally, some heavy traffic tramps across this surface. The flagstones around the door stay rock-solid atop the stone cribbing. If there were only a wooden liner holding the lip of that hole, I would worry about the stability of the patio base.

The 2 inch door is solid enough to walk on, yet light enough to lift. After one heavy winter, the door and the patio still fit flush in a suface that is comfortable even in bare feet. The door still lifts without binding anywhere. Nothing froze. And the outside faucet has survived the hose tugging and water fights for the first summer in memory. (Photo 5-1)

The outdoor tap was a special problem. The usual tap comes straight through the wall. This one, however, came through underground, then up the outside of the wall. It needed some extra support. More than that, it needed protection. The cedar struts atop the crib not only hold the faucet in place, they protect it from wheelbarrows, feet, and other abuse. The struts were also planned to hold flower pots and the trailing vine (which seems to prefer the shutter).

Drainage, too, was an unusual case. Because of the bedrock, this foundation drain runs *above* the well intake. The crib, as illustrated, extends under the drain. That's the hard way. Most intake pipes would enter above the foundation drain, and the crib would be a much simpler box. (Figure 5-1)

Foundation drainage, regardless of depth, works better under gravel fill. The design calls for setting the flagstones directly on the drainage gravel adjacent to the wall. (see Figure 5-4).

The other drainage consideration is slope. Slope the entire surface of the patio away from the house.

The stone floor is the simplest part to plan. It rests directly on the sandy soil. The area and pattern of the surface are completely flexible. The only part that is straight by necessity is the line along the door. Start here with straight-edged stones, then work your way out into the lawn until you run out of patience.

Locate the planting beds as camouflage. The little lilac bush will grow up to hide the meter box. The clematis, before it declared its independence, was meant to trail over the faucet. The big bed is there because it's the first spot to thaw. Crocuses and daffodils bloom here when the rest of the garden is still under snow. The sun on the wall behind, and warming the patio stones, makes spring come early to this corner.

Materials

Concrete, for footings
Anchor bolts
2 inch lumber, for cribbing and door
Galvanized nails
2 inch rigid foam
Foundation coating
Hinges, for door
Gravel
Sand, or sandy soil
Flagstones

The concrete goes in the footings. I used about half a cubic yard. Such a small quantity can be mixed at home, in the wheelbarrow if you can't get a little mixer (see Chapter 14).

Cribbing lumber should be rot resistant. I used local, rough-sawn cedar. Pressure treated lumber would last a little longer, but I was reluctant to use chemically treated wood around the well. This box, 6

Photo 5-1   The trap door to the well was a necessary but complicated feature of our patio design.

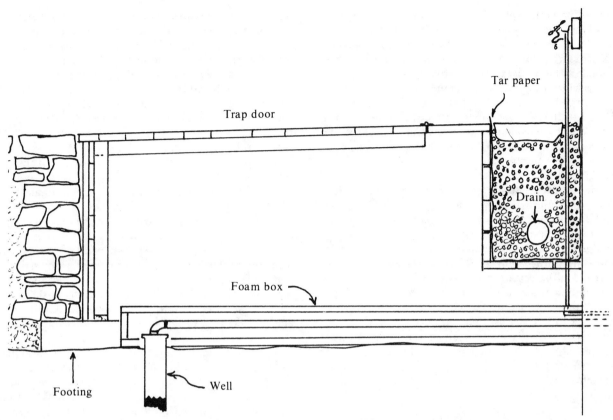

Trap door

Tar paper

Drain

Foam box

Footing

Well

Figure 5-1   My patio had special problems because of the location of the well and the placement of the tap on the outside wall. These created drainage problems requiring a special design.

feet × 3 feet × 3 feet (inside dimension), took 42 linear feet of 2 × 6's.

There are several types of rigid foam, sold by brand name. Ask your supplier for a foam that may be used underground. I used two 4 × 8 sheets around the crib, and a third sheet to box in the pipe.

Foundation coating won't really keep ground water out of the crib, not in the long run. It does, however, slow down the worms and the wood eaters. You can use anything similar that you might have on hand: tar, driveway sealer, etc. One gallon will be plenty.

The gravel is for backfill over the foundation drains. It helps, but is not essential. If the foundation drains are in good shape, and well-covered, leave them alone.

The sand makes an ideal bed for the flagstones. It compacts well, drains easily, and can be washed under the stones for leveling. The alternative is a gravel bed. Try to avoid using loam or clay under the stones. Loam compresses and clay doesn't drain. The amount of sand needed depends on the uniformity of the stones. A 6 inch layer should be lots.

Select the stones for their flatness, and for size.

Flatness is an obvious plus, especially in a barefoot area like this one. The thickness of the stone doesn't matter much ... thin enough to carry, and thick enough not to break when walked upon. But the surface area of the stone does matter. The broader the stone, the more stable it will be.

Preparation

The footings should be below the frost line, or on bedrock, whichever comes first. Here, it's bedrock, so we take extra measures with insulation. It keeps the water pipe from freezing, and it keeps the footings from heaving.

Scrape the bottom of the excavation clean. Brush off the rock, or scrape the subsoil down to a hard, undisturbed bottom.

Next, box in the water pipe with rigid foam. Cut the sheets with a kitchen knife and assemble the box around the pipe. You can glue the bottom pieces and the sides together with foam adhesive, but leave the top free to lift out when you need access to the pipe.

The order of assembly is important. If you look at the drawing (Figure 5-1), you'll see that the inner pieces are set in to brace the sides of the box, horizontally. Without the inner bracing, the box would collapse when you pour the concrete around it.

Form the outer perimeter of the footings with 1 inch lumber. I used 1 × 6 boards, not because the footing had to be that thick to support the crib, but to level out a very uneven bottom in the excavation. Similarly, the 24 inch width of footing either side of the water pipe was not for support so much as for tidying up the bottom of the hole. In fact, the only structural purpose of the footing is to hold the crib in place. A smaller footing would suffice for that. (Figure 5-2)

Level the top of the form. The foam box around the water pipe will probably not be level; it has to follow the pipe from the well to the foundation, and this is level only by accident. Do, however, level the outside form. And be certain that no part of the foam box is lower than the outside form. It won't matter if the concrete flows against the sides of the foam box — we've braced it against that — but we don't want the concrete flowing over the top of the foam. If that happened, the pipe would really be hard to get out, summer or winter!

Before you pour in the concrete, brace the foam box

in place with some heavy stones. Otherwise, the moving concrete might lift the box and separate the panels.

Now pour in the concrete. Move it around the forms with a spade, chopping it to flow the mix like lava. When the form is full, making a fat "U" of concrete around the pipe box, strike the excess concrete off the top of the forms. You can level the surface further with a float if you wish, but there's no point in troweling the finish. "level" is helpful; "smooth" is redundant.

In half an hour or so, when the water disappears from the surface, you can put in the anchor bolts. Remove the nuts and washers first, and grease the threaded ends of the bolts. This makes it easier to clean the concrete spatters out of the threads, and it slows down the rust.

Align the bolts where the wooden crib will sit. You want three bolts on each side, centered along the sill plates. There should be a bolt near each end of a sill, and another near the middle. Not exactly at the ends and middle, because upright studs will be nailed ends and middle, and we don't want the bolts to interfere with the studs.

The hooked end of the bolt, of course, goes into the concrete, with about 3 inches of thread left sticking out of the surface. Tap the bolt into place with a hammer. The vibration eases the stones aside and settles the wet concrete back around the bolt. Smooth the concrete around the bolts, check their alignments, then leave the whole thing alone for a couple of days. If you try to set the sill plates too soon, you could pull the bolts out of the uncured concrete. Leave it alone and keep it damp.

Wooden Crib

Start with a sill plate. Lay a 2 × 4 on the footing, beside the anchor bolts. Then use a small square to mark the location of each bolt on the board. Drill ¾ inch holes for the bolts, and test the board for fit. It should slip easily over the bolts. Put identifying marks on each plate so you can tell later which side is supposed to be up and which end is which.

If your excavation is big enough, you could assemble the crib in place. I dug a stingy hole, however, and didn't leave enough room to swing a hammer. So, I nailed three upright studs to each sill plate, then sheathed them with 2 inch cedar planks. I lowered each of these side panels into the hole and slipped the pre-drilled sills over the bolts. Replace the washers and nuts and tighten the sills into place. With both sides set up, I nailed the short lengths of sheathing across the ends.

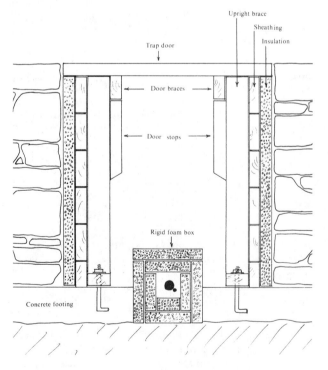

Figure 5-2   The trap had to be structurally solid before the concrete could be poured around it.

The sheathing at the ends brace the box against collapse. Assume, therefore, that if one end abuts the house you can leave that end open. You can omit the sheathing, but only if you include some kind of lateral brace at the top. (Figure 5-3)

Nail one or two planks across the top to hold the hinges. Normally, one makes a door and then mounts it on the hinges. I found it easier, however, to build this door backwards. I attached one board to the hinges, then nailed two long 2 × 4's to the underside of that board. the 2 × 4's would be the frame of the door. So, at this point, I had the frame (with its first sheathing board) hung on the hinges. The frame still had lots of flexibility. I lowered this skeleton door onto the crib and reached through the frame to adjust all the supports. The position of the door supports would determine the slope on the surface of the door. (Photo 5-2)

The object is to build in a slight slope to the side, and a slight slope away from the house, so the top of the door would shed water easily. I moved the supports, or door stops, up and down until the level said "tilt" in the right direction. When the two slopes were right, I nailed the supports in place. The supports held

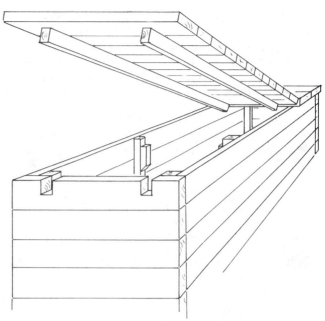

Figure 5-3   The crib box is braced well together with lateral sheating boards.

Photo 5-2   The door supports must be adjusted to achieve a level door surface which will be flush with the patio.

61

the 2 × 4's in exactly the right position for slope and closure. Now it was a simple matter to nail sheathing across the 2 × 4's, and trim off the ends. By assembling the trapdoor in place, we ensured that it would rest evenly on all its supports, close snugly, and slope in all the right directions.

The final step for the wooden crib is to tar the outside and add another shell of insulating foam. The tar goes on with an old whisk broom or a very stiff paint brush. Swab it into all the cracks. Start at the top and work you way down the side, brushing out the dribbles and allowing the waste to pool along the bottom edge, where the crib meets the footing.

Cut the foam panels to fit and stick them right onto the tar. The top of the foam should be flush with the top of the crib, and should fit under the lip of the trapdoor.

Stone Crib

Start the stone wall on the concrete footing. And be generous with the stone — it should not have to rely on the wooden crib for support. A wall 3 feet high will need a base at least 2 feet wide.

Lay the stones on the flat, not standing on edge. The task will be a great deal easier if you lay it out course by course, using stones of about the same thickness in each course.

Set the first stones by the box, butting the straightest edges against the foam. Then work towards the back of the wall, fitting the stones like a flat mosaic. The important thing is to watch the heights. Try to keep the tops of the stones fairly even, and avoid leaving high spots in the middle of the course. If most of the stones in the course are 6 inches thick, a stone that sticks up 8 inches will be a sure source of grief in the next course. Fill that awkward spot in the 6 inch course with two 3 inch stones, and save the big one for a place where it won't make a hump. (Photo 5-3)

As you finish each course, fill the smaller gaps with stone chips, sand or a shovelful of soil, brushed into the cracks. This filler helps to stabilize the wall. If a stone does wobble, the filler trickles into the gaps and lodges the wobbler more tightly.

You can take the initial wobbles out of the wall as you lay each stone in place. Press down on the stone at its edges. It may tip back and forth. Tip it towards the side that makes it most level. That leaves a gap under the other side. Shove a thin stone into the gap. That shim in the gap should hold the bigger rock steady. If it still wobbles in another direction, try another shim.

Underground, and against the box, this wall is not subject to much abuse. You can shim it any way that

Photo 5-3   *As you are building up the wall, keep it level by using stone chips to fill in the gaps.*

takes out the wobbles. But, for the record, the best way to shim is from behind. The dirt behind the wall holds the shim in place. And that's the next step.

As you finish each course, fill in behind it with dirt. It is better to backfill in thin layers like this, tramping the fill as you work. When you get to the top of the wall, the fill will be compacted, leaving a firm base for the patio floor. The alternative, backfilling the whole at the end of the job, is more likely to leave soft spots. By the following Spring, the soft spots will be low spots in the floor.

Lay each course in the same order: start at the face, then fill in stones behind. Keep the tops as even as you can, and shim the wobblers until they're steady. Fill the cracks, then fill the dirt in behind.

There are two other rules for the building of any stone wall. First, bridge the cracks in the course below. This staggers the joints and strengthens the wall. Secondly, tie the front face of the wall to the back by including an occasional stone that spans from front to back. In other words, turn a large stone at right angles to the crib, so it runs *across* the wall rather than along it.

The tricky part is to come out even on top. The top course on the wall is also part of the patio floor, and it must fit flush with the trapdoor. Save the heaviest, flattest stones for last. Measure the thicknesses of these top stones, and compensate for that thickness on the wall. In other words, if the capstone course is 6 inches thick, build up the wall to within 7 inches of the top of the door. (The extra inch is for wobbles and shims.) If the stones are less uniform than that, leave even more space for the top course. You can, after all, raise the caps to the proper level by adding thinner stones underneath. Measure the top course, then allow for that thickness in the height of the wall.

The cap course must not only fit snugly along the edges of the door, and flush with the door's top, but the stones must also be leveled for the floor.

Floor

The floor, in truth, is not "leveled". It is sloped gently away from the house. Because this patio fits into a corner, it slopes away from two walls. The trick is to establish a perimeter of reference points to define the slope.

We already have one central reference point in the door. You will recall that we sloped the door as we built it in place. That slope was a 1 inch drop in 8 feet. I laid an 8 foot board atop the door and a level on top of the board; I had to raise one end of the board an inch in order to make it level. Nothing very technical in that.

The next step is to mark the house. Lay the board along the door and slide it up until it touches the house. Mark the house where the underside of the board touches. That's the floor level at that point.

Now extend the mark along the walls of the house. Use the board, the level and the ruler. Level the board across the first mark and into the corner. Raise the end in the corner so the board slopes 1 inch in 8 feet. Mark the corner. Next, mark the other wall. This time start at the corner and drop the board 1 inch in 8 feet. Mark the wall 8 feet from the corner.

At this point we have one mark in the corner and one mark on each wall. Now snap a chalk line from the corner and across one of the wall marks. The chalk line marks the entire edge of the floor along that wall. Do the same on the other wall.

The outer perimeter works exactly the same way; except there we'll have to use stakes instead of a mark on the wall. Decide, roughly, where you want the outside corner to be. Drive a stake there. Set one end of the 8 foot board on the trap door, the other end on the stake. Drive down the stake until the board slopes 1 inch in its 8 foot length. You can add any number of intermediate stakes, or you can extend the perimeter beyond eight feet by setting a series of stakes. Any series of stakes can be checked with a taut string; the tops of the stakes should all coincide with the string.

The stakes, the door, and the marks on the wall, together define a plane that slopes uniformly in two directions. We want the surface of each flagstone to coincide with this plane. Again we'll use the long board between any two reference points. Then align the stone with the underside of the board.

Start with the capstones around the trapdoor. The front edges of these stones are right. Their alignment with the top of the door establishes that. But their back edges may be too high or too low. Set the board across the top of the stone, with one end resting on the door and the other end resting on the stake (or the wall mark). The surface of the stone should be flush with the underside of the board. If the stone is too low, shim it up higher with thin stones underneath. If the stone is too high, remove some material from underneath.

Now fit a perimeter of stones along the wall of the house. If you've excavated down to the drainage bed, fill this area with clean gravel, then set the stones directly on the gravel. Align the edge of the stone with the chalk line on the house. Then level the stone as before: set the board across the top, with one end on the door and the other end at the chalk mark. Raise the stone or lower it by adding or substracting gravel in the base. (Figure 5-4)

Leveling on the sand fill is even easier. Again, the board is the reference. If the stone is too high, tip it up and examine the sand bed. The stone made an impression where it rested. These are the high points. Rake away the high points and the stone will settle lower (and more solidly) into the bed.

If the stone is too low, raise the edge with a crowbar, dump sand at the gap, and wash the sand under the stone with a blast from the water hose. With a crowbar, a hose, and a barrow of sand, you can level a

Photo 5-4   Leveling a sand base stone floor is easily accomplished with a crowbar and a hose.

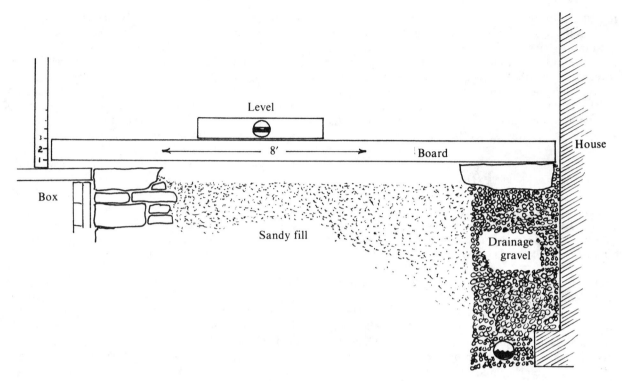

Figure 5-4   Make sure the stones on each side of the hole are level with each other by aligning with a level and a long piece of board.

flagstone floor in minutes. (Photo 5-4)

Fitting the contours of the stones, edge to edge, can be as fussy as you want it to be. It's possible to fill the gaps with mortar, as we did for the porch, but I wanted the patio to be more adaptable than that. I'm happy to let the grass grow between the stones, so the fit wasn't critical. I settled them in without cutting, letting nature's contours come as close as they could.

The hardest stone to fit in a floor is always the last one. The last stone has to fit a hole instead of an edge, so it has to be the right shape all the way around. For this reason, we started the patio at the edges and worked inwards, towards the planting areas. The contours around the planting areas aren't planned. That's just how the stones happened to end.

If the planting doesn't work out, it's easy enough to change things around. Lift a stone, replace the sand bed with some planting soil, and you have a planter. (Photo 5-5)

Alternatives

Not everyone has a well to protect, but the patio hole can be adapted to a variety of other uses. Just the way it is, it would make a fine little root cellar. It never freezes. The bottom is cool and damp all year. The upper part is cool and dry. Carrots, apples, potatoes and cabbage would keep all winter in a box like this.

Many homeowners have gone back to the rain barrel, storing soft, unchlorinated water for hair washing or gardening. The patio hole could be a cistern, or underground rain barrel. For this purpose, I would eliminate the wooden crib liner and make the hole circular. Lay up the stone wall in a circle, mortaring the stones this time. Then plaster the face (the lining of the cistern) by trowelling on a layer of cement-rich mortar. When the lining is dry, swab it with swimming pool paint, or with foundation coating. We made a cistern like this eight years ago and it's still in perfect shape, without a leak.

The function of the hole determines the best construction method. The wood and stone crib shown here is probably the most elaborate approach. That's partly the trauma of that pre-Christmas dig, and partly my aversion to the concrete block. If you don't find them offensive, consider a block crib for the hole. You would still need a footing, and some reinforcement (eg: steel rods in the footing, extending into concrete-filled block cores), but a solid block crib would make as good a hole as the stone does. And you wouldn't necessarily need the wooden liner. Concrete in the

cores can hold bolts, which could hold the fixed part of the top. Hinge the door to this and the door would fit just as snugly as mine.

Or, eliminate the hole entirely, and make a patio around you own household. It may not be fancy for company, but it certainly does tidy up that tangle of hoses, cords and mud.

Photo 5-5   Container gardening on a stone patio is as simple as lifting out a stone wherever you want to plant.

Chapter 6

Table

I should begin, in self defense, by admitting to a deep aversion to things that blow away in the wind. That, and a very large dog named Dickens who has a twenty pound tail that can knock over chairs and upset an entire table full of drinks with a particularly enthusiastic wag.

So, when we discussed the need for a table by the barbeque, Dickens' tail figured largely in the planning (he gets particularly enthusiastic at barbeques).

Fortunately, few people are owned by such dogs and, consequently, there may be a limited appeal for this 2000 pound table. But building the table did illustrate some useful techniques for moving very large stones, techniques that every stonebuilder can use, regardless of what you're building.

Consider this: the top slab shown in the photos weighs approximately 1600 pounds. I found it by the roadside, almost half a mile away. Fighting back temptations to use neighborhood tractors, power winches and four wheel drives, I tried to make it as hard as possible, giving myself no more mechanical help than the average homeowner might find at the tool rental shop.

I did recruit Doug and Dan, but Doug was feeling unwell that day, and Dan had just emerged from a winter in the darkroom. Physically, we were about as awesome as the rusty old truck we used. The three of us, plus the truck, were no match for Dickens. The question was whether we could beat the rock.

Levers

The first trick was to pry the thing out of the ground. Here, there is no substitute for a stout steel bar. Not the little hooked crowbar that bent on a stubborn nail, but the kind they call "railroad" bars, or "stone" bars. We used a regular 4 foot bar from the hardware store, plus a longer piece of steel junk that might have been a buggy axle once upon a time.

You will remember from high school physics that a lever needs a pivot point, or fulcrum. The closer the fulcrum is to the rock, the bigger the rock that can be lifted. Unfortunately, moving the fulcrum closer also reduces the range within which the weight can be raised. Never mind. With rocks this big, you only need one. Accept the fact that you can only move it a few inches at a time.

Use a smaller rock as a fulcrum. Shove the fulcrum in close to the slab. Then jab the lever down between them and pry. With two bars prying at once, you double the advantage. In many cases, however, you will find it easier to use the bars alternately — one lifts and holds while the other changes position.

To move a slab like this horizontally, along the ground, raise it far enough to slip one of the bars underneath, then pull *up* on the end of the bar. This reverses the leverage. The rock itself becomes the fulcrum. The mechanical advantage remains the same. That is, the length and position of the bar determines how much weight you can move.

In our case, there wasn't much hope of moving a 1600 pound slab very far with these relatively short levers. Even with the longer bar, we could barely raise the stone high enough to roll some smaller rocks beneath it.

Chains

With the slab raised just a little off the ground, we could pass a chain beneath it. (Photo 6-1)

There are two common hooks used with heavy

Photo 6-1   With the hugh rock raised just a little off the ground, there was enough clearance to pass a chain beneath it.

chain: a grab hook, and a slip hook. The grab hook has a narrow slot, wide enough for a single link, but too narrow to let the next link pass. The slip hook is rounded and the chain slides though it easily. The slip hook end of the chain goes around the load. When the pull begins, the chain will tighten like a noose around the rock.

The chain always goes *over* the hook, pinching it against the rock. As the chain tightens, the pressure is on the hook to stay in place. (Photo 6-2)

Photo 6-2   The chain goes over the hook, so when tightened, it will hold firm in place.

At this point, the easy thing to do would be to hook the other end of the chain to the truck and drag the rock to its final resting place. For a short haul, on dirt, with a decent clutch, that's the logical course. Dragging on a road, however, is hard on the road and on the chain. This was borrowed chain. And we had, after all, set out to do it the hard way. No dragging.

Raising the rock onto the truck was the challenge. Part of the answer was a ramp. (Photo 6-3)

Photo 6-3   Getting the tabletop onto the truck was a challenge requiring a ramp which we constructed with available materials.

Ramps and Rollers

This particular truck would seem naked without at least one big plank in the back. Most rocks too big to lift can be rolled, or slid, up the makeshift ramp and into the truck. Back at the job site, the ramp is used again to get the rock to the top of whatever we happen to be building.

The tabletop rock, however, was more than muscle power could move uphill. The obstacles were sheer weight and friction.

Friction can be combatted with rollers. It's possible to use two, but better with three: one front, one centre, and one to move forward as the load advances. In this case, we used a length of well pipe, and two small cedar logs. With logs, be sure to lop off any knots or protrusions that might impede the rolling.

Weight was a bigger problem. We had access to a chain hoist, and a load binder. The chain hoist would have been more convenient since it allows a longer pull before re-positioning is necessary. But load binders are cheaper to buy, cheaper to rent, and likely to be more easily available to casual users.

A load binder cinches chains a notch or two tighter. Truckers use them to tighten the chains that hold the load on the truck.

Here, one chain goes around the rock. Another is attached to the truck (actually to a log, braced ahead of the wheel wells). The load binder hooks between the two chains and drawns them together.

With the slip hook around the rock, and the grab hooks on the binder, we were ready to inch the stone up the ramp. Doug strained at the handle of the binder and was barely able to budge the rock. He slipped a length of pipe over the handle, making it longer by at least two feet. This type of handle works like a lever — the longer it is, the less effort needed from the human helpers. With the impromptu pipe extension, Doug could pull it without gritting his teeth.

Each pull on the binder moves the rock a link or two closer. (Photo 6-4) Unhook the chain, take up the slack, and re-hook it that bit closer. It is slow, but we've got all morning and all of us are past the age when such things must be hurried.

When the slab finally arrives at the bottom of the ramp, there are two more little problems to overcome. Solving the first problem creates the second.

The first problem is friction. The rock wants to drag the ramp along with it. The solution is simple: lever the rock up a few inches and put some rollers underneath it. Less friction, so the ramp stays put and the rock rolls up it.

Unfortunately, the rock now rolls down the ramp even more easily than it rolls up it. So the few inches grained by each pull on the binder are more than lost when we release the tension to rehook the chain. That's the second problem. The solution is equally simple, but considerably more strenuous. After each little gain, two of us hold the rock in place with the bars, while the third man releases the binder and re-hooks it.

It is at this point, when the father of all rocks is poised halfway up to the truck, when the only thing holding it in place are two bars scrabbling for a toehold in the dirt below, that I think of two more important questions: First, is everybody wearing steel-toed boots? And, second, if it starts to go, which way shall I

Photo 6-4   Moving up the ramp was a slow process — an inch at a time.

jump? There is no question of trying to catch the beast if it does start to slide, or if the ramp starts to break. Under the law of gravity, this is no petty misdemeanour. This is your major felonious rock hanging there over my precious toes. Remember: steel-toed boots, and be ready to get the heck out of the way.

Finally, Peter the Rock rolls over the hump and into the bed of the truck. (Photo 6-5) Doug cinched it up as far as it would come with the load binder, and we chained it down where it sat, ready for the tense ride back to its intended place by the barbeque.

Tense? Yes. A teen-aged, rust-ridden truck that was only good for half a ton when it was new, was now all aquiver under the 1600 pound behemoth. Six years ago we had to put a sheet of plywood in the back because too many people were falling through the bed. Atlas it ain't.

Photo 6-5 *Finally, onto the truck where it was cinched down on top of logs.*

Preparing the Base

Meanwhile, back at the barbeque, a base was ready to support the slab. The principle was that of the three-legged stool. A tripod cannot wobble — no matter what. This was to be a three-legged table. Actually, the "legs" are boulders, odd-shaped lumps of granite that survived ten years in a stonemason's yard because they were too mis-shaped to fit into any wall, and too cussed hard to cut into anything nicer. I'd have thrown them out years ago if they had not been too heavy to throw.

We began, as usual, by digging a hole. Not too deep. This table, despite the weight, actually rests on a "floating" pad. The hole had to be deep enough to provide some drainage, and to hide the concrete pad. I dug down through the grass roots to some harder subsoil. The hole was about 12 inches deep annd 3 feet across.

Drainage is a 6 inch bed of gravel, dumped in the hole, spread out evenly, and tamped down hard with the end of a 4 × 4 post. That leaves the hole 6 inches deep. We wanted the table top to be 22 inches above the ground. The slab was about 8 inches thick. So the "legs" had to stand about 20 inches high (6 plus 22 minus 8 equals 20).

Finding legs was a simple matter of rolling around boulders until we found three that would stand the right height. And, as long as there was gravel in the bottom of the hole, the legs would be adjustable. Anything close to 20 inches would be fine.

Moving the boulders to the hole gives us another excuse to talk about the lazy man's way of moving rocks. As usual, the rocks we wanted were found too far away to roll them to the hole. And nowhere near where we could load them into the truck. The wheelbarrow was made for times like these.

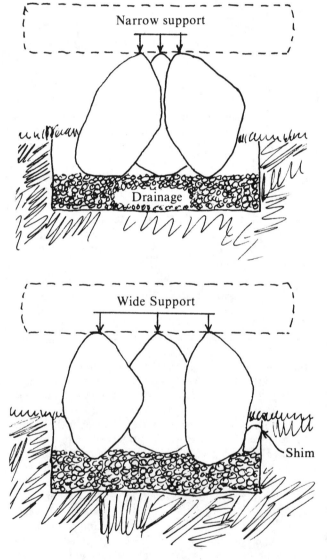

Figure 6-1 *Try to arrange to support rocks in such a way as to provide a wide support for the tabletop, as in the lower drawing.*

The big advantage of the barrow is being able to load it with a much bigger stone than you could hope to lift. Tip the barrow over on its side. Roll the rock into the pan. Then tip the barrow upright again. Liz, who barely weighs 100 pounds after a big meal, hauled much of the 60 tons of stone for the house this way.

You can use the same tip and roll technique to load those little single-axle trailers that are often used behind garden tractors.

We dumped the boulders into the hole, then wiggled them around until they stood up unassisted — 14 inches higher than the edge of the hole. The slab would rest on the three high points of the boulders. What really matters, then, is the position of those three high points. We want the three points on a more or less level plane. And we want them spread, as much as possible, away from the centre of the slab. If the three high points were clustered towards the centre, the three-legged table would become more like a one-legged table. (Figure 6-1)

Finally, try to position the boulders for temporary stability. Lowering a 1600 pound slab on this arrangement would be unsettling if the rocks were inclined to roll. We shimmed one boulder against the side of the hole. The other two propped one another up.

Permanent stability comes with the concrete that will surround the three boulders, locking them into position. Though it is tempting to pour the concrete now, stabilizing the base before the big top descends, resist the temptation. The bottom of the slab is unlikely to be perfectly flat and perfectly parallel with the top. If that's the case, leveling the three high points will not guarantee a level top. The idea is to get everything together, then leveled, and then committed for posterity in concrete.

Now the trick is to lower the big slab onto the base as slowly and carefully as you can. First, build a temporary crib around the base. Use concrete blocks, timbers, anything solid that can be stacked higher than the top of the three boulder legs. The object is to roll the slab onto the crib, then lower the crib piece-by-piece until the slab comes to rest on the boulders.

When we had a crib (of sorts) arranged, we bridged the gap between truck and crib with the same two planks that had been the loading ramp. The crib was high enough that the plank bridge was virtually level. A downhill run would have made unloading easier, but much less accurate. *Photo 6-6*

Photo 6-6   Getting the tablestop off the truck wasn't any quicker as we had to inch it down slowly to the ground.

71

Getting It Down

The two bars levered the slab back onto its rollers, then levered the rollers onto the ramp. Slowly and gently, we nudged the slab out onto the planks and over the crib. This is the time to rotate the slab if necessary, lining it up for final positioning.

Now we needed some independent fulcrums for the bars: concrete blocks, chunks of firewood, solid pieces of varying heights. You could raise the slab by using the crib as a fulcrum, but then you can't remove that bit of crib. The pry points have to be independent of the supports that are being removed.

First, raise the slab a corner at a time and take out the rollers. Now the stone is resting on the ramps. So raise the stone a corner at a time and take out the ramps. Now it's resting on the crib. This is why we wanted fulcrums of varying height. Each time we take a bit out, the stone is that bit lower.

Dismantling the crib is a tricky business. The slab tips one way, then another. If it tips too far and slides off the crib, or collapses the crib, you have to start all over again (possibly with a cast on your foot). We used one bar to raise the slab, one bar to brace it from the opposite side; and the lucky leftover worker got to put his hand underneath and pull out the next piece of cribbing. If it is hard to control the slab at this stage, an automobile jack or two would be handy for lowering by degrees while the cribbing is adjusted.

Photo 6-8   *In place — only an earthquake could move it now.*

When it finally rested on the boulders, we breathed three premature sighs of relief. Premature because the level said it wasn't. We figured out which of the three boulders was the low corner, estimated how much higher it had to be, and raised the slab again. Not much, though — we didn't want to tip it off at this stage.

Independent fulcrums and two bars held up the side of the slab while I reached under and re-positioned the boulder leg. Since it had to be raised, I rolled it aside and laid more gravel on the bed.(Photo 6-7) The boulder went back to its original position, albeit a little bit higher, and we lowered the slab back onto the base. This time the level said "perfect".

We poured concrete around the base of boulders, filling the hole to ground level. When the concrete was set, so was the table. The sheer mass of the thing keeps it in place. There is no need to mortar it to the legs. Not even Dickens' tail could move it now. (Photo 6-8)

Alternatives

The point, we agreed, was to demonstrate that it is possible to move big rocks around. You don't have to hire a crane, or leave some monster stone in the front yard to be whipper-snipped around forever. The three of us put the one ton table together in about three hours, with common hand tools and five dollars worth of concrete.

Photo 6-7   *To level the tabletop we had to use fulcrums to lift up the slab and reposition the legs.*

Moving a big rock is one thing. Making something useful of it is another. Not everybody can find a big, flat slab in the neighborhood. You can, however, make them flat. Many boulders have a splittable grain. Look for the striations. If the lines run straight, or better still, if you can see straight cracks in a boulder, open one with a chisel and steel wedges.

If a rock is almost — but not quite — flat enough to be useful, a concrete grinder will take off the high spots and smooth the surface. These grinders are normally used to smooth a concrete floor, or to finish terrazo. You can rent a flat grinder, and buy the concrete heads in many large hardware stores that cater to the construction and masonry trades.

Chapter 7

Entryway

The approach to Dan Maruska's lower entrance suffered from every ill that nature could inflict on a hole. Tree roots, unstable sandy soil, and a steep slope all pushed at the high side. Rainwater filled it. Ice heaved it.

The hole's only virtue was a magnificent view. The very fact that a hole could have a view made it worth the face lift, and serves to illustrate just how steep is the slope.

The first attempt to make an entrance in the hillside involved a concrete block retaining wall. The hill pushed the wall into the hole. Obviously, nature didn't want a hole right there. Preserving the entrance called for stronger measures.

Fixing the problems that afflicted Dan's hole gave us a chance to consider all the wrong ways, and the right ways, to build a retaining wall. If you want a simple retaining wall, check the chapter on planters. This one if for the really tough cases.

Design

Building a better whole means solving the drainage problems, the frost problems, and most of all it means holding back the pressure of all that dirt that would really rather be in the hole. (And you thought it was bad at the office!)

There are three basic tactics for combatting the pressure behind the wall: the curve, the taper, and the inside brace.

The *curve* is elegant in its power and simplicity. Look at the photo of the long wall, the one that holds back the high side of the hill. The hill pushes against the outside of the curve. The pressure has to straighten out the curve before it can break through the wall. In

effect, it must push a 20 foot wall through a 15 foot gap. It is the same principle as an arch, or a dam.

The original plan for the cellar entrance had called for a simple, straight-sided hole. the straight wall (a habit afflicting carpenters and cement block masons) is the same length as the gap it fills. So even a little pressure can push it over or bulge it out of place. Stone masons, happily, have broken the straight line habit. The first, and most important part of the new design was to erase the straight wall and draw a gentle curve, a curve that aimed its bow right back at the strongest force opposing it. That alone solved most of the problem.

The *taper* refers to the shape of the wall. Seen in profile, both the face and the back of the wall lean in, from a wide base to a narrower top. This is to resist the force that would tip the wall over. Stability depends on the width of the base, and the height of the centre of gravity. Sloping the faces lowers the centre of gravity and widens the base. Most of the weight, in other words, is concentrated near the bottom of the wall. To tip that mass over, the hill must first raise the weight of the wall beyond the edge of the base. It's like tipping over a pyramid.

The *inside brace*, in this case, is the set of steps. The hole wants to collapse in on itself. The long, uphill wall is arched to resist that pressure, but the short wall curves the wrong way. It starts with a built-in bulge. Granted, the pressure is much less on the downslope side, but even the little pressures can win in the end if you don't put up some resistance.

The steps hold the two walls apart. The dirt behind the short wall can collapse it only by pushing both walls uphill, or by crushing the concrete steps between the two walls. Neither outcome is likely.

The curve, the taper, and the inside brace sound

more like baseball than masonry. And, like baseball, there is more intuition and artistry to it than technical sophistication. We drew the curves with the toe of a boot, and tapered the faces according to the rule of "whatever looks right".

One of the joys of working in stone is the rough and ready character of the material. The bumps and wiggles are part of its appeal. So why try to straighten them out? For the mason, a curve is easier to build than a line. It's easier to roll the big boulders into the bottom course than to lift them onto the wall later. and pouring steps from wall to wall is easier than trying to form the ends. If these shortcuts also happen to coincide with sound principles like the arch, the pyramid, and the inside brace, so much the better.

The design, however, had to consider more than the sheer weight of dirt behind the walls. Frost and drainage cause other problems that exacerbate the pressures. To some extent, frost and drainage are related. It is the water in the soil that freezes. The more water retained in the soil, the more serious are the frost problems.

We planned the drainage to remove water from behind the walls, and from beneath the floor. In addition, we had to remove rainwater that would collect above the floor, and we had to accomodate existing foundation drainage.

It sounds complicated, but there is really only one simple principle to understanding your drains: water runs downhill. The existing drains around the foundation gave us starting points. All we had to do was ensure that every additional section of new drainage began at these levels or lower, then sloped away from the house and down the hill.

The floor is a special problem. Obviously, it had to be drained. Less obvious, in the summer heat, was the eventual need to clear out the snow. Snow insulates. Where snow is removed, frost penetrates farther and faster. That's why driveways and sidewalks crack and heave, while the lawn seems undisturbed by frost. We could protect these retaining wall from frost by burying the footings under earth and snow. But the floor would be shoveled bare. So, the design included a frost-stopper: a layer of rigid foam insulation panels under the floor.

The foam is a recent, though hardly original, solution to the problem of frost beneath the floor. Highway builders in cold climates now pave on foam insulation for the very same reason.

Materials

Quarry waste
Cement
Sand and gravel
Floor drain
Drain elbow (3 inch)
Drain pipe (8 foot length of 3 inch diameter)
Flexible drain pipe (25 foot length of 4 inch, perforated)
Foam insulation panels (2 inches thick)

Quarry waste is one solution for stonebuilders who don't have stones. Dan's hillside lot is pebbled with the odd little granite boulder, but that would hardly do for something so grand as this. We calculated the volume of the walls, divided by 150 pounds per cubic foot, and concluded we would need about 15 tons of usable rock — more than Dan could carry in the back of the station wagon. So he found a limestone quarry that would sell rough rubble as it was blasted out of the bedrock. (Quarries are listed in the Yellow Pages). Stone, loading, and forty miles of trucking came to $200.

Better quality stone might have resulted had we had time to pick through the piles before they were loaded. The big, yellow scooper-upper wasn't that discriminating, and the delivery included plenty of gravel-sized chips and some back-buster blocks that had to be broken up before we could move them. (See Chapter 13 for advice on cutting big stones into usable ones).

Rough and ready as our estimates were, the result was close. When the job was done, there were two decent stones leftover, and enough of the gravelly chips to provide drainage under the steps.

The quantity of sand and cement required will depend on the quality of the stone. Rough stone, with wide joints, takes more mortar than flat-sided rocks would need. The quarry waste used here was about as knobbly and irregular as stone can be, and the consequences was a bill for forty bags of cement (including the concrete for footings and floor).

Using rough stone, and making your own concrete, count on one small truckload each of sand and gravel. The sand goes into the concrete and the mortar. The gravel goes into the concrete and the drainage channels.

For a big job, where materials arrive by the truck load, give some thought to where the piles go. The messy part is the mixing, and washing out the mixer. So put the mixer downhill from the sandpile, and downhill from the building. The sandpile, for convenience, is close to the mixer. The gravel is beside the sand (with boards between them to prevent fraternization and unnecessary screening). The rock pile, if possible, belongs on the other side of the work site. That keeps the mixer wash from staining the stones. And, in this case, it meant that the rock pile was higher than the hole in the ground, so the heaviest blocks could be rolled *down* ramps and onto the wall.

Foam insulations have one quirk that will necessitate a private chat with your building supply dealer. It seems that some brands taste better than others. Termites and rodents relish the stuff. Ask for advice on the type of foam that is least appealing to your local pests.

Preparation

Dan had the rough excavation done with a backhoe. Bill and Ed finished the job with shovels. The shape of the hole was set by the curve of the walls we had planned. The depth of the hole would vary.

Where the footings went, under the walls, we wanted a depth of at least 5 feet. That would put the footing below the local frost line (your building inspector can give you a depth for local conditions).

When we got to the frost line, however, a little probing indicated a much harder layer just a few inches lower. We had hoped for bedrock, but found hardpan shale instead. It's compact, and more than adequate to hold this wall with no subsidence. It took more concrete to put the footings that bit deeper, but the extra security was worth it.

Getting down to hardpan also brought the footings below those soft upper layers of organic soil, where roots and worms and rotting vegetation can shift the walls. In general, footings should be below the frost line, below the darker, organically active soil, and below any visible roots. If you hit bedrock on the way to that depth, count your blessings and stop right there.

The proper width for footings depends on soil type and the weight to be borne. In essence, the footing distributes the weight of the wall over a wider area, supporting the weight in the same way that a snowshoe supports the weight of a walker. For a house, this is a critical consideration, and the "Stonebuilder's Primer" covers the subject in more detail than is necessary here. Here, a wall 5 feet high, 3 feet wide at the bottom and 2 feet wide at the top, bears down on the base with 625 pounds per square foot of base area. Clay, one of the weakest soils, will support 1500 pounds per square foot, more than twice the load we had designed for this footing.

So Ed and Bill dug a trench for each wall, curving away from the house, following the hardpan bottom. The trenches were 3 feet wide, and always more than 5 feet deep. They scraped the bottoms flat, and cut the sides as cleanly as they could — so we could use the trenches as forms.

Between the trenches, where the floor would go, we excavated to a level 15 inches lower than the adjacent

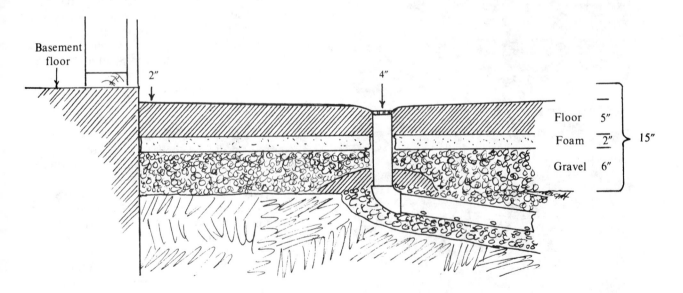

Figure 7-1   Slope the surface toward the drain in the center of the walkway.

basement floor. Here's why: on top of the dirt, we would need a 6 inch layer of drainage gravel, 2 inches of foam insulation, and 5 inches of concrete floor. That's a total of 13 inches of material. We added two extra inches for slope — so the entrance floor would drain away from the basement door. That's a 15 inch excavation for the floor.

The final bit of spade work involved the floor drain. Bill dug a narrow trench from the centre of the future floor to the downhill side of the pit. The bottom of the trench, itself, sloped downhill. Since the sides of the excavation were unstable sand, we installed the drain pipe right away, before Bill's trench re-filled itself. (Figure 7-1)

Gravel goes in the bottom of the drain trench, then the pipe. Drain pipe comes with small holes along one side. Lay it with the holes facing *down*. Then cover the pipe with 6 inches of clean gravel — no sand! (Photo 7-1) The upper end of the pipe (the end that will connect to the floor drain itself) should be protected from dirt and heavy feet. Plug it with a rag, and cover it, temporarily, with boards.

Photo 7-1 Place the drain pipe down first, then cover with 6 inches of clean gravel.

Finally, before you start the concrete mixer, scrape out the trenches where the footings go, removing all the loose dirt. The footings should rest on compact, undisturbed soil. Loose dirt would eventually settle, leaving a void under the footing. If you must fill low spots in the excavation before the concrete is poured, fill them with rocks, then stomp on the rocks to make sure they won't sink any lower.

Pour the footings with concrete (see Chapter 14). If time or materials are short, this is the place to call for the ready-mix concrete truck. Calculate the volume of the footing trenches, or forms, and order by the cubic yard. If you're intent on mixing your own, prepare for a long day and a short list of helpers. You'll need someone to mix, one to wheel the stuff back and forth, and one to work it into the forms. Do accept the necessity of pouring at least one entire side at a time. Stopping the pour at the end of the day, to resume the next, leaves a weak spot in the footing. One weak spot won't wreck a job like this, but it's a bad building habit nonetheless.

As the concrete goes into the form, work it with a shovel, churning up and down to settle the goop into all the corners and eliminate the air pockets. This chopping motion is a far better way to move wet concrete around than using a rake, which can separate the aggregates. Chopping it around until it flows, glacier fashion, is easier on the worker and the best way to treat the concrete.

Dan's footings, remember, were a little deeper than we had planned, and took more concrete than had been budgeted for. We padded the mix by dropping in large stone as we worked. We were prudent enough to use only the most misshapen lumps, and lazy enough to use the biggest ones down inside the concrete, where they would stay exactly where they fell and did not have to be wrestled into any particular position.

When the trenches were full, the footings were done. But, while the concrete was still soft, we set the first course of wall stones directly onto the footings. We aligned the faces along the curve, and filled the vertical spaces between the stones with mortar. This early marriage of footing and wall saved us one bed of mortar that would have been needed to start the wall after the concrete had set. More importantly, it bonds the wall to the footing in a strong, direct fashion. Now, the only way the hill could push the wall off its footing would be to break the entire first course of limestone blocks. Breaking a simple mortar joint is easy compared to the effort that would be needed to break a solid array of 200 pound blocks, burried half in the footing and half in the wall.

Floor

The lower portions of the two walls will act as a form for the floor. Pouring in the concrete is easy. The hard part is to get the levels right, so that water runs down the drain and not into the basement.

Start with the drain itself. Assemble the drain and the elbow and slip it onto the end of the pipe. There will be considerable movement in the assembly, since the pipe is held only by the gravel around it. Move it up and down until the top of the drain is at the desired level. We set Dan's drain about 4 inches lower than the basement floor. If your level isn't long enough to check the hieght of the drain, extend it with a long, straight board.

When the drain is at the right height, dump some gravel around the elbow to hold it there. Then level the top of the drain, wiggling it back and forth until the top is perfectly flat. When both height and level are right, dump a bucket of wet concrete around the assembly. (Photo 7-2)

Now you can use the drain as the primary reference point. From here, you can set all other floor levels: gravel, foam, and concrete.

First, use bright chalk or crayon to mark the level of the finished floor. Run a level from the top of the drain to the stone wall (the stone we set atop the footing). Mark the stones 2 inches higher. If we can finish the concrete floor at those marks, the entire perimeter of the floor will be 2 inches higher than the drain, and 2 inches lower than the adjacent basement floor. The rain should go down the right hole.

Photo 7-2   *When you have placed the drain in position and it is level, dump a bucket of concrete around it to hold in position.*

With marks all around the perimeter, it's a simple task to dump in the gravel and rake it smooth, making an under-floor drainage bed about 6 inches deep. We nailed a short board to a long handle and used it to tamp the surface, settling the gravel and leveling the top at the same time. When the gravel bed is smooth and hard, recheck the elevations to ensure that you still have 7 inches to the finish mark (2 inches of foam and a 5 inch floor).

Cut the foam boards with a kitchen knife, shaping the edges to fit the contours of the hole. As you set each piece in place, check to be sure there are no voids beneath the foam — low spots in the gravel base.

Pour the concrete right on the foam. (Photo 7-3)

Photo 7-3   *Pour the concrete right on top of the foam boards.*

The edges of the sheets have a nasty habit of rising up as the weight goes on them someplace else. If this happens, you'll have a miserable time trying to stuff the foam back into place, especially if a bit of concrete gets underneath. Stand on the sheets. Weight them down at the edges with the first load of concrete. Then pour in the rest of the mix.

Work the concrete into place with a shovel, just as you did the footings. Don't let it cover the drain. Stuff a rag in the drain to keep out any overflow of cement.

When you're filled to the finish line, level the top — roughly — with a straight board. Set the board on edge and use a worker at either end, "sawing" the board back and forth across the surface. A wave of wet concrete will rise in front of the board. Use the board to work this wave of excess away from the drain and towards the edges of the floor. Go back and forth across the floor several times, until all the high spots have been leveled off. Don't try to smooth the surface at this stage. Just level it.

Chances are that the swishing back and forth has left a puddle of soupy cement around the drain, the lowest spot in the floor. Now, aren't you glad you stuffed a rag in there? I was. Scoop out the watery stuff and throw it away.

While the wet surface is firming up a bit, use the time to make a long-handled "float". The business end is a smooth board, 18 to 30 inches long, and 6 to 8 inches wide. Cut a 45 degree bevel on the end of a long 2 × 2, or a broom handle, and nail the beveled end to the smooth board. Add a couple of short, diagonal braces to complete the assembly. Hammer down the ends of any nails that might be sticking through the bottom of the board. (Figure 7-2)

Figure 7-2   You can easily make a concrete "float" for smoothing the concrete with leftover lumber.

Use the float like a shuffleboard stick, going back and forth over the floor in long, smooth strokes. With a little practice, you'll learn to tip the handle up and down to keep the leading edge of the board from digging into the surface.

The float levels the floor, and smooths it by bringing more sand and water to the surface. This floor, of course, should not be flat, but should slope slightly towards the drain. So, float the excess away from the drain and towards the perimeter.

The float won't provide the final finish. For that, you'll have to wait several hours until the surface dries out some more. How long, depends on the weather and on how wet the mix was to start with. Usually, it's ready to trowel about the time you're ready to go to bed. That's when it's hard enough to walk on, and soft enough to leave a shallow footprint.

Use the flat, steel trowel like a mini-float, sweeping

it over the surface with moderate pressure until a skim of water comes to the surface. After the edge of the trowel digs up a few stones, you'll learn to tip it just enough to bring the leading edge over the rough spots rather than through them.

Finally, remember to clean out the drain after every trowelling. When the surface around the drain is still soft, but no longer runny, pull out the rag stopper and replace it with the cast iron drain cover.

After the final trowelling, cover the floor with plastic. The longer you can keep it damp, the harder it will be.

There's a psychological milestone here — a point at which discouragement turns to satisfaction, and horrified shudders of the tidy-minded start to turn to smiles. Until now, each day's work has left a deeper hole and a bigger mess than the one we started with. And the toilers knew that every effort was destined to be buried out of sight beneath the final product. After the floor, the hole gets smaller and neater by the hour. And every improvement will show at the end.

Walls

The wall is the essential, most visible part of the entrance. It is, at the same time, one of the easiest parts to build. There are a few simple rules to make it strong, and a few more to make it easier:

1. *Keep a straight face.*   Remember that the earth will be doing its best to push the wall over. Don't help it by building in a lean in the wrong direction. We want to slope the face back slightly towards the centre of the wall, tapering towards the top. A lean in that direction makes the wall stronger. Leaning out over unsupported space makes the wall weaker. Perfectly vertical would take a plumb bob, batter boards, or a better eye than mine. All we needed was to stay on the safe side of vertical — leaning back just a little. So as each big stone was set in place, we used the level to check the vertical alignment. Little stones we aligned with the big ones.

2. *And a level head,*   to make fitting easier for the mason. Lay the stones in courses, grouping stones of more or less equal height. If you start level, then lay a course of uniform height, the top of that course will also be level. Any stone the right height will fit on the wall.

Uniformity of height is not as hard as it sounds. Dan's 15 tons of limestone was about as randomly shaped as dynamite could make it. The pieces we could lift were mostly 6 to 12 inches thick. So, give or take half an inch, we had a couple of tons to choose from for any given course. For example, laying out a course

10 inches thick meant choosing from 2 tons of rock that happened to be between 9½ and 10½ inches thick. If everything we picked up seemed to be 6 inches thick, we threw that aside and made the next course 6 inches high.

3. *Stagger the vertical joints.* Any stone the right height will fit, *but* it should not end on a joint in the course below. The mortar joints are the weakest part of the wall. If you line them up, you give the wall a place to crack. If you stagger them, the wall has to break rocks in order to crack straight. Find a stone the right height, then make sure it ends short of the joint below, or beyond it. Bridge the gaps, in other words.

4. *Start at the face.* Most of the stones will lie flat on the wall, not standing "on edge". It's more stable that way, and it's easier to build in uniform courses (see Chapter 13). So the face of a rock, the exposed part, might be the straightest edge of a slab. The straightest edge is exposed and the wigglety-jigglety backsides get hidden in the middle of the wall. Begin each course at the face, lining up a series of those nice, straight edges.

5. *Then the back.* Fill in the back as if it were a face that will never be seen. Straight edges and pretty facets don't matter. Height does. So does bridging joints. The back should be about the same height as the front, and solid. Forget pretty.

6. *And fill in the middle.* Stuff anything in the gaps between the face and the back. Use bigger rocks if you can — it's faster and will require less mortar. But don't waste any nice, square blocks as filler. Don't even waste a straight edge as filler. Get rid of the round rocks and the ugly ones by jamming them in these holes. One caution: don't let the middle build up in high points, higher than the face or the back. The result will be a rocker in the next course, a wobbler. If the heights won't match, leave the middle just a little lower than the face or the back.

That completes the course. Work some mix between the stones, slater a new bed of mortar on top, and start the next course at the face again. (Photo 7-4)

7. *Bond the face to the back.* The first six steps describe a two-faced wall, joined in the middle with a filler of mortar and rubble. Restore its integrity from

Photo 7-4   Build up the wall by placing in the facing stones first, then fill in the middle and back with irregular material.

time to time by adding a big stone that reaches from front face to back, or by overlapping two large stones by at least 6 inches. For a project like this, it's enough to remember that every seventh or eighth stone in the face should overlap the back of the wall.

8. *Pick the capstones.* In any wall like this, the best stones go on top. If possible, they should also be the biggest stones. There are two reasons: First, heavy stones on top help hold the wall together. They weigh down the little rocks below them. The top course takes the most wear and tear. Lawnmowers bump it, kids balance on it, snowshovels scrape it, and there is nothing above the capstones to weigh them down. Except for a piddley little bit of mortar, the only thing holding them in place is their own weight. So the capstones should be heavy! The other reason to put the big ones on top? They look more impressive up there.

The critical point, however, is to remember the capstones before you get there. As we sorted through the rockpile, course by course, we set aside a series of the biggest slabs. By the time the wall was half-built, we had enough of these toppers to make a capstone course. Most of them were about 12 inches thick. We knew ahead of time, then, that the last course would be 12 inches high. We knew how high the finished wall should be, so we aimed for a line 13 inches lower than that (12 inches for the capstones and another inch for bumps and mortar). The ends of the two walls that joined the house would match the slope of the land. The outer ends, the ends that held the steps, would be level with one another. By picking the capstones ahead of time, we came within half an inch of every desired elevation, without the help of line or transit.

That's all: a straight face, an even course, a solid back, and fill in the middle. Stagger the joints and bond across the top when you remember. Each course is a mosaic of stones. Cover it with mortar and start the next course.

More on Drains

As the wall went up, the earth came down and filled the space behind the wall. When Bill got tired of digging it out, we put in a bed of gravel behind the wall. We joined the 4 inch flexible pipe to the old foundation drain, and snaked the new pipe around the retaining wall to the point where it would eventually disgorge the rain — downhill of the house and beyond the steps. (Photo 7-5)

Bill filled the low spots under the pipe with more gravel, raking and shoveling the gravel until the pipe ran downhill all the way. Then he covered the pipe with another 6 inches of gravel, and covered that with strips of tarpaper and plastic (to keep the dirt from filtering into the drains).

With the drains covered, the dirt could fall in as fast as it wanted. We went back to building wall, with one man working at the face and the other man working at the back of the wall. The man behind the wall trampled the earth as it continued to cave in, compacting the fill as he worked.

Steps

The steps came last. The reference point was the floor. Each step had to rise 7 inches, so it was a simple matter of fitting a 7 inch board across the floor, cutting it to fit snugly from one wall to the other. Each face form after that was likewise fitted wall to wall, and checked with the level before bracing it in place. (Photo 7-6)

Photo 7-5   Using flexible pipe for drainage is an easy way to take care of awkward situations.

The hole, of course, was somewhat deeper than the 7 inch face forms. Dan filled most of the hole with the rubble left from the quarry stone, and with the leftover gravel. He left 6 inches of free space beneath each face board, so the concrete would be 6 inches deep at the back of each step and 13 inches at the face. The rubble and gravel provide a firm base and lots of drainage. Extending the steps from wall to wall locks them in place so they cannot subside.

Mix the concrete, pour it into the forms, and finish the surface just as you did for the floor. One exception:

Photo 7-6   The steps come last. Place the concrete forms in position from one side to the other.

the floor had only a top surface to finish. A step has a hidden surface — the vertical face behind the form. Work the concrete into this space with some vigour. Chop out all the air pockets. Then tap the form board with a hammer — not hard, but thoroughly. This vibrates the mix behind the board and brings water and sand to the fore, filling any little holes around the gravel and leaving a smooth, vertical face.

If you have an edging trowel, or can borrow one, run the tool along the inside edge of the form at the finish troweling stage. This rounds the lip slightly, and smooths it.

If you can't get an edging trowel, slide a regular trowel along the inside of the form, near the top. It won't work as well as an edger, but it's better than leaving a rough lip.

Ed and Bill and I left Dan to finish the steps on his own. (Photo 7-7) The few little stones that were left, he could lift on his own. And we'd done enough to earn invitations to the annual lamb roast.

Photo 7-7   Done! All that remains is to remove the wood forms after the concrete has dried.

83

Chapter 8

Porch

Somewhere out there is a lady who should know that this porch is dedicated to her. She slid off the road one dark winter night, stuck to the scuppers in a back country drift. She trudged for nearly a mile through the wind and snow before she saw the light in our front window. Heartened, she left the road, stumbled through the trees, and waded across the dark, drifted lawn towards the light.

There was, indeed, a door beside the lighted window, but the threshold was 3 feet high. Worse, when she did reach up to knock, her knuckles hit nothing but the plastic stapled over the frame. By the time we heard her, and waved her around to the back door, the lady was half frozen and close to tears. All she wanted was to use the phone, she whimpered through chattering teeth.

We brought her in and toasted her in front of the fireplace for awhile. And then we explained, as gently as we could, that we didn't have a phone.

And so, cold lady — wherever you are — if you ever get stuck again, there's a real door now, steps to reach it easily, and a roof to keep off the snow. We've even removed the plastic and installed a phone.

Design

Other than making it easier to get in the house on a cold, snowy night, this porch displays a variety of useful construction techniques. The three principle problems were support, drainage, and careful planning of elevations.

For a wooden porch, or a concrete pad, footings are less important. If the supporting soil moves a little with frost or subsidence, the wooden floor can flex and readjust with no harm done. A concrete pad can rise and tilt as a single, unbroken unit. But a stone structure of this size can't move and flex without cracking mortar joints. The answer is to plant the underpinnings on immovable footings.

Not counting earthquakes, the main subterranean movers and shakers include:

Frost — soil expands as it freezes, contracts as it thaws.

Roots — expand as they grow, heaving everything around them.

Decomposition — organic matter shrinks as it rots.

Subsidence — soil compacts under pressure.

The footing, in order to beat these forces, must be designed for depth and width.

How deep? Deeper than the frost. Deeper than the soft, organic soils. Away from the fat surface roots of nearby trees. The municipal building inspector can tell you how deep the frost gets in your own area. This "frost line", however, is a general notion. Consider the specifics of your own design. The ground may freeze more deeply under walks and steps, where the insulating snow is cleared away. On a sunny, protected side of the house, frost penetration will be much less.

How wide? The wider the footing, the more weight it can bear without compressing the soil and sinking. The footing distributes the weight of the structure over a wider area, in the same way that a snowshoe distributes the weight of a walker. Calculate the weight of the structure, and consider the type of soil.

Compact sand can carry 3000 pounds per square foot. Compact silt is rated at 2000 pounds per square foot, firm clay at 1500 pounds per square foot. A low porch like this one would be within the limits, even without a footing. But we widened the base a little

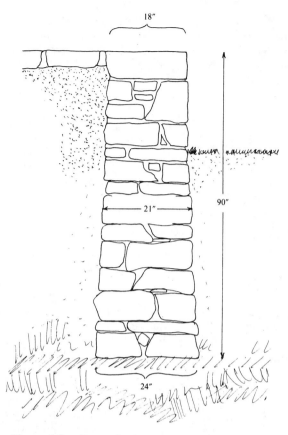

Figure 8-1 For calculating the weight per square foot of the wall, measure the height, average width and width of the base.

anyway — just to be on the safe side. (Figure 8-1)

My calculation for the deepest, heaviest part of the wall went like this:

$$\frac{Height \times Average\ width \times Weight\ of\ stone\ per\ cubic\ foot}{width\ of\ the\ base}$$

$$\frac{7.5ft \times 1.75ft \times 150\ lbs/cu\,ft}{2ft} = 984\ pounds\ per\ square\ foot$$

That's about one third the limit for the compact, sandy subsoil under that part of the wall.

That part of the excavation was dirt — all the way down. In other areas, we hit bedrock. Where the bedrock was close to the surface, we planned the best and simplest of all foundations: solid concrete, poured in an earthen trench.

The problem with concrete is the cost. In an ankle-deep trench, concrete is an easy choice. When you're up to your armpits before you hit rock, it takes a lot of concrete to fill the hole. Here, in the deep sections, we

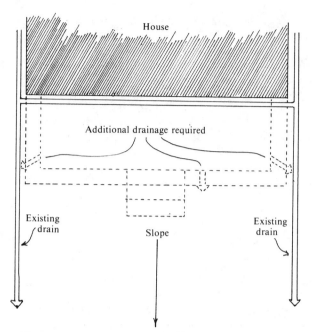

Figure 8-2 Plan for drainage by placing the drain to allow water to escape downslope.

laid dry stone foundations.

The dry foundation — flat stones, laid without mortar — is a fast and easy base if you have plenty of stone. Laid stone on stone, it won't subside so long as the base is wide enough. And it's unlikely to heave so long as the base is deep enough (below the frost), and so long as we can keep it dry.

"Dry" is one of the keys to a solid porch. Good drainage is as important as good footings. The reason is simple: it is the water in soil that freezes and gets pushy. The first consideration is the existing drainage around the house foundation; we can't allow the new structure to interfere with existing drainage. The second consideration is water under the porch. Assume that, no matter how we build, some water will get under the porch. Then plan drainage exists to let the water out. (Figure 8-2)

The dry stone foundation is self-draining. Water simply runs out between the stones. The concrete foundation, however, can act like a dam, holding water behind it. This is of special concern where the concrete is poured on bedrock; water can't go under this footing, so we have to provide an exit through the footing.

The perimeter wall, above ground, acts like a retaining wall for the fill inside the porch. Normally, the weakest point in a retaining wall is the centre of a long, straight section. The front of the porch is a long, straight, retaining wall. So, we planned the steps in the centre, at the weakest point — like an outside brace holding back the fill.

The floor stones rest on compacted sand. The key word is "compact". You must take out all potential subsidence before laying the floor. Sand compacts more easily than regular dirt, and it drains quickly in case any water comes through the floor. And, we'll reduce the chance of water coming through the floor by mortaring the flagstone joints, and by sloping the floor away from the house.

Planning the steps takes a little more public input. The building code, and possibly the inspector, should have a say. No one is worried about the steps falling down. They're worried about visitors falling down the steps. The step's height and depth (called the "rise" and the "run") are carefully limited. So is the handrail requirement. And, when we're starting with rough stone, safety should also include some judgement as to the uniformity of the surface.

The elevations of steps and floor have to be planned together. Start with the desirable floor level, then measure the distance to the ground. The building code will specify minimum and maximum rise for each step, and will probably also insist that each step have the same rise. So you have to get from the floor to the ground with uniform steps that fall within the allowed range. The bottom step can't be three inches or twelve to make up the difference. You can't have half steps. The division has to come out even. If you can't make the floor elevation with legal steps, you will have to change the floor level, or change the ground level.

Materials

Stone
Sand and gravel fill
Sand and cement, for mortar
Concrete, for footings
Drain tiles or pipe
Gravel, for drainage beds

The quality of stone is critical for the floor, and for the steps. Get the flattest, smoothest stone you can find. If you can't get enough flat material, use concrete steps and floor (see "Alternatives"). Any sort of stone will do for walls and foundation, but sedimentary slabs will be faster to lay.

Fill for under the floor should be sand, or sand and gravel. To estimate quantities, calculate the volume of the space (length × width × depth). Then add enough to replace the top layer of sod and soft dirt. Waste rock, or rubble, can be used as fill, but you should still have sand to wash into a rock fill. Otherwise, the sand bed under the flagstones will trickle away into the fill.

You'll need sand and cement for mortar between the flagstones, and mortar in the wall. Most of it goes in the wall. I ordered a truckload of sand, even though the wall took only a fraction of that. Most of the sand went into the fill, after the wall was done. I saved a small part of the pile to finish the floor. Nothing was wasted.

Concrete is optional. I used it in one of the footings because the excavation was very shallow, and because the ready-mix truck was nearby another job anyway. Concrete is also useful for capping a dry stone foundation. A concrete cap, just below ground level, can level the top of the stone base and give it added stability. It isn't essential, but if you have the concrete, use it here.

Drain tiles, or pipe, may be needed to cross a solid, concrete foundation, or to replace old drains disturbed by the excavations. Don't bother ordering any until you've finished digging and checked the condition of existing drains. A couple of tiles, or a short scrap of plastic pipe, will be enough to span a concrete foundation.

Drains are bedded in gravel. (Photo 8-1) If digging disturbs the beds, you'll need fresh gravel to restore the

Photo 8-1   Bed the drain pipe in gravel.

drains when you backfill the excavation. Also, you can improve the drainage through a dry stone foundation by backfilling on either side of it with gravel. If you need only small amounts of drainage gravel, you can substitute rubble or cutting chips from the main stone pile.

Preparation

Mercifully, it won't be necessary to excavate the entire area of the porch. Start with the perimeter walls, and the area under the steps. If there's a chance that bedrock is near enough to warrant a concrete footing, lay out the dig precisely — with string. Then cut the sod out cleanly with a spade, leaving a sharp, firm edge to the trench.

Dig down through the dark layer of topsoil, roots and worms, down to bedrock or the frostline, whichever comes first. Scrape the loose dirt off the bottom, and clean right to the edge of the trench. The bottom should be flat and hard. If there are low spots, don't fill them with dirt; scrape the rest a little lower to match, or pack the low spot with stones.

If you intend to pour concrete on rock, clean the rock surface with a wire brush. And cut the sides of the trench cleanly, leaving the base a little wider than the top.

Wherever you pour concrete on rock, consider the natural drainage. The section I wanted to pour ran across the slope of the lawn, and across the apparent slope of the bedrock. The concrete would act like a dam, holding water on the upslope side — inside the porch! So, we included a drain at the lowest dip in the bedrock. Two tiles, end to end, spanned the bottom of the trench. I covered the ends of the tiles with rocks to keep concrete from filling the drain. And, for the same reason, I covered the joint between the tiles with plastic.

The existing house drainage is even more important. Watch for the old gravel beds as you dig. It's sometimes possible to dig past a drain, holding the gravel bed and the tiles in place with a well-braced sheet of plywood. But any gravel that does get mixed with the dirt in the new digging should be removed. The dirt would wash into the drains and eventually clog them.

Poured Foundation

If you have cleaned the trench, brushed off the bedrock, and protected the drains, you can pour in concrete right to the top of the trench. Work out the air pockets with a spade, chopping and kneading as the mix is poured into place.

You can extend the concrete by dropping large rocks into the partly filled trench. I find this a convenient way to get rid of those big, useless boulders that won't fit anyplace else. Wet the rocks first.

There's no need to smooth the top of the pour. You can wait until it's hard, then mortar the first course of stone atop the concrete surface. Or, set the first course of stone directly into the still-soft concrete. The latter method requires being ready ahead of time with all the stone at hand. Otherwise, the concrete will set faster than you can fit the stones. But setting the first course in concrete does save one mortar bed, and it makes a strong bond between the stone wall and the foundation.

Stone Foundation

Start with a mosaic of flat stones. Put the largest, flattest ones at the sides of the trench, just as if you were building a two-faced wall. Stomp on the stones to settle them on the dirt bottom. Then fill any gaps between the two faces with smaller stones. Don't let

Photo 8-2   Set wobbly stones on a firmer footing by adding small stones or shims in small openings.

this centre fill rise higher than the faces, or the next course will be harder to fit.

Start the second course at the sides again, making two faces with the edges of the larger stones. Lay the stones on the flat, not "on edge", and lay a few of them across the wall, spanning the whole width if you can. Try to cover the joints in the bottom course, so each stone overlaps at least two below it. Bridging the cracks, or joints, like that is what gives the wall its strength.

Wobbly stones can be fixed with small, flat "shims". (Photo 8-2) Collect a little pile of stone chips and scraps, flat enough to skip on water. Put a big stone on the wall and try to wobble it back and forth. When you hold it down on one side, there will be a gap under the opposite side. Stuff a shim into the gap. Now try to wobble it again. If it still wobbles, adjust the shim or shove in a thicker one. If you have a choice, always shim under an edge that will be buried in the centre of the wall. A shim at the face can work loose; a shim in the centre will be jammed in position forever.

Lay each course in the same order: on the flat, start at the face, bridge the cracks below, fill the centre gaps. If you have some gravel around, finish the course with a shovelful of gravel. Brush it down into the cracks to help jam the bigger stones in place.

If the trench is wider than the wall, fill the gap on either side with gravel as the wall goes up. (Photo 8-3) This helps the drainage around the wall, and keeps soil out of the joints between the stones — soil that could freeze and expand.

Also, restore the house drainage as the foundation wall rises past it. Replace the gravel, and ensure that the tiles (or pipe) are aligned, and not plugged with silt. Correct any dips, or backslope, where water might collect.

As the foundation wall approaches the top of the trench, choose between a concrete cap, and mortaring the wall directly atop the dry stone base. (Figure 8-3) If you've used large, flat stones that cross the top of the foundation, or at least reach well back into the centre of the wall, then you really don't need concrete. The top of the stone base should already be flat, and level, and well tied together. If the stones are smaller, or rounder, a concrete cap may help bond it all together and stabilize it before you begin the mortared wall.

I capped the base of the steps with concrete, partly to level an uneven base, and because the steps are

Photo 8-3 *Fill up the gap between the wall and trench with loose gravel.*

Photo 8-4 *To achieve a stable platform for the steps it is a good idea to pour a concrete base or "cap".*

89

subject to greater stress than any other part of the porch.

To cap, form the top edge with 1 inch boards, nailed at the corners and braced into place. (Photo 8-4) Because my base was uneven, I had to shape the bottoms of these boards to conform to the rough contours of the stone. I didn't need a tight fit, but I did want the form to be stable and level. Five minutes with an axe is all it takes.

Wet the stones, work in the concrete, and leave it to cure.

Walls

The critical part of these walls is getting the elevations exactly right. First, mark the desired floor level on the side of the house, from one end of the porch to the other. In order to shed water, the floor must slope slightly away from that mark.

This porch would have a slope of 2 inches across the 8 floot floor. In other words, the floor will touch the mark along the house, and will drop 2 inches lower than that at the front corners.

Begin the masonry at the end walls. Slap a thick bed

of mortar atop the foundation and lay in a course of stone. By now, the routine should be familiar: first the face, then the back, then fill in the gaps between. Stagger the vertical joints by overlapping each course, and tie the wall front to back by laying some stones across the wall, spanning the whole width of the wall.

Insofar as possible, match the thickness of stones within a course. Matched stones, set side by side, will be easy to span with another stone across the top. Keep the front face and the back face at about the same height. And don't let the rubble in the centre rise higher than the faces.

When a course has been fitted, work mortar into the joints from above. Then spread a new bed of mortar and start the next course.

Keep the corner rising vertically by setting each cornerstone with plumb bob or level. I used the corner of the foundation as a reference point, and plumbed each cornerstone from that point.

A wall is normally capped with a course of heavy stones to keep all the little stones in place. If the capstone is heavy enough, gravity will hold the wall together. Before the end wall gets too close to the top, select the stones for the cap course. These caps will top

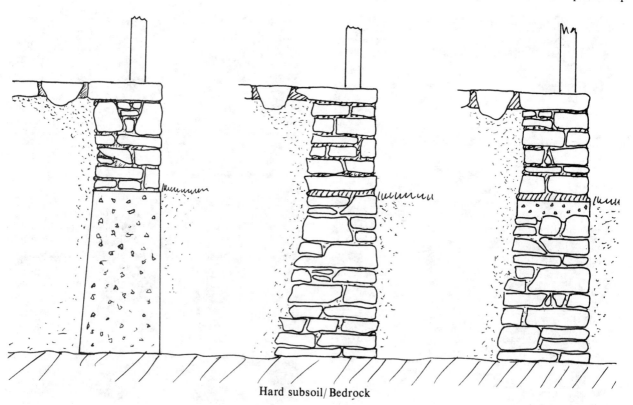

Hard subsoil/Bedrock

Figure 8-3 You have your choice of three types of foundation walls: a concrete wall (left), dry wall (center), and a concrete cap on dry wall (right).

the wall *and* form part of the floor. They must, therefore, be flat on top for the sake of the floor, and heavy for the sake of the wall. Finally, for appearance's sake, it's better if the capstone course shows a uniform thickness at the face.

Select the capstones, then lay them out in some approximation of a fit. You can do the fine fitting later, when you actually put the stones on the wall. For now, all we need to know is how much height to allow for, and where. Measure the maximum thickness of a stone. Add half an inch or more for mortar. The wall must end just that far short of the final elevation. If the capstones are all the same thickness, you have only one dimension to allow for. If the caps vary in thickness, the wall must be stepped to match that contour.

It isn't easy to build up to some imaginary line in space. As a guide, set an elevation stake at the corner. Drive in the stake until its top is at the final corner elevation (2 inches lower than the mark on the house). Stretch a string or a straight board from the top of the stake to the mark on the house. This line represents the final floor level, including the slope. Now you can build up the wall until the top is the proper distance from the line, a distance that will just accomodate the capstones. (Figure 8-4)

Lay the capstones on the end wall. Align the tops with the elevation guideline (or guideboard). That will give them the correct elevation and the correct slope. Then level each stone in the other direction, along the porch. Set the level on top of the stone, and adjust the stone in the mortar bed until it's level in one direction, and sloped in the other direction.

Once you have the first two capstones set, one at the corner and one against the house, leveling the ones in between is easier. Set a board across the first two caps, and align everything else with the underside of the board.

Don't bother filling the joints between capstones at this stage. There will be lots of time for that later, when the rest of the floor is pointed.

When both end walls are finished to floor level, stretch a string from one cornerstone to the other. This string defines the front edge of the floor. All the rest of the capstones can be aligned with the string, elevations for the steps will be measured from the string, and the face of the front wall will be plumbed from the string.

The remaining wall, along the front, is built just the way we built the end walls. It's a little easier, because now we have two corners to hold the guideline. And it's a little slower, because this wall incorporates the steps.

Steps and wall are built together. Start with a step, then fill in wall arount it.

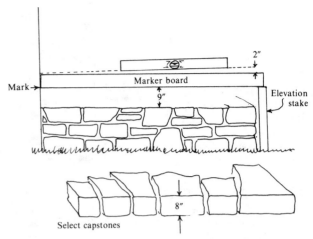

Select capstones

Figure 8-4   Building a level structure by eye isn't easy. The simple use of elevation stakes and string will make your job easier.

Steps

Again, elevations are the key and the starting point. Place four stakes near the corners of the base pad for the steps. The object is to set the tops of these stakes at the exact elevation for the first step. (Figure 8-5)

The top of this first step had to be 19½ inches below the string (three risers, each 6½ inches high). I drove in the two back stakes until they were each 19½ inches from the string. Then I leveled the two front stakes from these. The tops of the four stakes defined the top of the steps. A board, placed across the tops of any two stakes, would give me a surface guide.

Start at the front of the bottom step with the flattest stones you can find. When you have enough stone selected to surface the step, measure the thicknesses, and raise the base to accomodate them — just as we did for the capstones.

Level each stone with the board across the elevation stakes. Check it across the front, back to front, and diagonally. When a second stone has been set and leveled beside the first, check to be certain that the two surfaces are flush with one another. Minor differences along the joint can be corrected with pointing mortar, or chipped away with the chisel. Major differences probably mean that the stone is not flat enough. Replace it.

Start at the front and work back towards the wall. Once past the line where the next step begins, you can quit being so careful with the leveling. Keep it roughly level, as you would for the top of any course, but don't worry about the fine fitting of surfaces. Finish the course back into the wall. Then fill in the wall to the same height. (Photo 8-5) Leave the first step to set

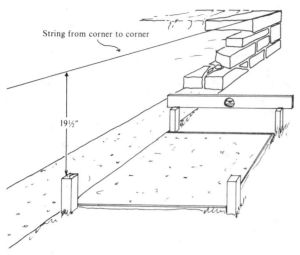

String from corner to corner

19½"

Figure 8-5 Use stakes in each corner of your step area, and measure down from the string to keep the proper height of the steps in check.

Figure 8-6 The third step was aligned with the wall, and the floor would rise above it one layer of stone.

Photo 8-5

overnight before you begin the second one.

Mark a chalk line across the first step, where the second step begins. We'll set up the front rank of stones along that line. This time, though, we'll do the elevations a little differently. We could, if we wanted, reset the stakes 6½ inches higher. But now we're one step closer to the wall, and we have the bottom step as a reference point. We can measure *up* 6½ inches from the bottom step, or *down* 13 inches from the string.

Set the front rank 6½ inches high with the ruler. Adjust the top surfaces with the level. And check them, one against the other, with a straight edge. Once again, as you work you way back past the point where the next step would begin, you can be more carefree with the surface. And again, fill in the wall to this new height. (Photo 8-6)

The third step, in this design, is aligned with the rest of the wall. It is built as another course on the wall. Just be certain, when you get to the step area, that the surfaces are flat, and exactly 6½ inches higher than the last step (or 6½ inches lower than the string).

The top step overlaps the wall a little, but most of it rests on the fill, like the rest of the floor. So we'll treat it like floor.

Floor

The first part of the floor is the capstone course. You've already capped the end walls. Now cap the front wall in the same way: select the stones, measure their thicknesses, build up the wall to accomodate those thicknesses under the string. (Photo 8-7) Use the string as the guide for both elevation and alignment along the face. For narrow front caps, you can estimate the slope. For example, a ¼ inch drop across a 12 inch stone is the same slope as we set on the end wall (2 inches in 8 feet).

Larger caps should be checked for slope. Snap a chalk line along the side of the house, from one end wall to the other. Place a long, straight board across the top surface of a capstone. If one end of the board is flush across the top of the stone, the other end of the board should hit the chalk line. If it doesn't, adjust the slope of the stone.

That completes the perimeter. Now comes the all important task of filling the area under the floor.

Begin by removing sod, weed roots, and all organic matter. Where practical, dig right down to a harder, subsoil base. But, if the topsoil is too deep to bother, at least get down beneath the roots and worms. Remove the loose soil that would compact, and scrape out a flat, undisturbed bottom.

Set up a ramp for the wheelbarrow and start hauling in fill — sand, or mixed sand and gravel. Spread it around in thin layers, 3 or 4 inches deep, and compact it.

Better still, get the neighborhood kids to compact it. Look at the nearest playground — the soil is packed like concrete! You needs lots of water and pounding feet. Give the kids a hose and an hour's suspension of the rules. (Photo 8-8)

If feet aren't enough, give them a tamper: a length of timber, or a 2 × 2 with a small board nailed on the end, or a pogo stick, or stilts.

Don't tell the kids this, but the water isn't just for their amusement. It washes the sand down into all the little nooks and crannies. This is especially vital along the wall, where the dirt was excavated and backfilled

Photo 8-7   Begin laying the floor of the porch with a cap along the wall using the string as a guide.

Photo 8-8   *You must provide a solid base for the floor. Fill with sand, wet and it and pound it down. This is one job you can always get the kids to volunteer for.*

Photo 8-9   *Use a level to adjust each stone, aligning it with the mark on the house.*

with gravel. Encourage the help with stories of moats and war dances.

Continue filling in layers. Spread one layer out and compact it before you add the next. Bring the level up almost to the surface, leaving just enough room for the flagstones. Then, if you can, leave the fill for at least one good rain. The longer you can leave it, the more settling it can get out of its system.

When the fill has had enough water and kids and pogo sticks, spread a fresh layer of sand on top and set the biggest flagstone against the house. The biggest goes first because it's too big to lift up and down to fit against another stone; this is the one you only want to set once. I put it against the house and in front of the door for two reasons: That's where the traffic will be heaviest, and the big slab in the centre will give us an immovable reference point from which to level the rest of the floor.

Judge the thickness of the stone, then try to match the depth of the sand bed to it. Wet the sand. Then drop the stone into place. If it's really big, use the crowbar to shove it around until it's exactly where you want it.

Now level it. First, level it against the mark on the house, the one that shows where the floor ought to be. Then put the level across the top and level the stone along the length of the porch. (Photo 8-9) Finally, adjust the slab to the slope by using the long board —one end of the board resting on the stone, the other end on the perimeter wall. The board should sit flush on the surface of the stones, and hit the mark at the house.

Setting the stone — whether to the mark, to the level, or to the board — requires a fractional raising or lowering of the stone.

To lower it, lift the high edge and rake away some of the sand. Look at the wet sand under the rock. You'll see smooth impressions where the stone rested, rough sand where there were gaps. To lower the stone, you must rake away the spots where it rested. Drop the rock into position again. If it's still too high, lift it again, look at the impressions, and rake. (Photo 8-10)

To raise the stone, dump fresh sand around the edges. Then get the hose and the crowbar. Use the crowbar to raise an edge to the right level and hold it there. Then, with the hose, wash sand under the edge.

94

Photos 8-10   *To level the stones, you'll have to adjust the sand layer underneath. Roll back the stone and note where the rock was resting (left), and then to lower it you will need to rake away sand (right).*

(Photo 8-11) Ease off on the crowbar. The stone should maintain its level. If it settles below the mark, dump more sand, raise the stone, and wash the sand into the gap. You might have to repeat this process at several points around the edge before the stone stays in position.

With the big stone at the back, and a perimeter wall at the front, the rest is filling in gaps. The board crosses all the holes, marking the surface elevation and slope.

Fitting the mosaic, edge-to-edge, can be as fussy or as casual as you like. The casual way is to eyeball the angles around the face of a stone, then look for an opening along the edge of the mosaic where the angles match those on the stone. (Photo 8-12) The fussy way begins just like that, and ends by trimming the edges of the stone that almost fits to make it fit exactly (see Chapter 10).

The forgiveness factor in the casual fit is putting mortar between the stones. The mortar doesn't stick the stones together so much as it jams them into position, not giving them room to wriggle loose.

Wet the joints first. Then mix a batch of softer than

Photo 8-11   *To raise the level of a stone, lift the edge with a crowbar, dump sand along the edge, and wash it down beneath with a hose.*

Photo 8-12 If you are fussy about fitting angles closely together, this will require trimming off the edges with a chisel.

Photo 8-13 Use two trowels together when mortaring the joints to prevent smearing it on the surface of the stone.

usual mortar.

One way to get the mortar down into the joint without smearing it all over the surface is to use two large trowels in tandem. (Photo 8-13) Make a "V" of the two trowels, with the narrow bottom down in the joint. Load the mortar on one trowel and use the other one like a cleaver, chopping the mortar into the crack. Work it down as far into the joint as you can, until it's just slightly over full. Move along and fill the next section, until all the mortar is gone.

Don't try to smooth the mortar yet. Wait until the surface begins to stiffen. This might take as little as an hour on a hot day, longer in the shade or the cool weather.

Wet a small, rectangular trowel and slide it along the joint. Press down hard, forcing the mortar down to the level of the stones and bringing a skim of water to the surface. Pass over the same joint several times, until the mortar surface is slick and smooth. (Photo 8-14)

The trowel's passage may push up a ridge of excess mortar on either side. Cut away the excess with the edge of the trowel. Don't try to clean it up now — you'll only make a mess. Just slice it away from the good mortar and leave it. When it's dry, you can brush it off with a soft broom.

Cover the fresh mortar with plastic, to keep the moisture in. And avoid walking on the stones until the mortar has had three or four days to cure.

Alternatives

For many, the hardest part may be finding stones flat enough to make a floor and steps. I split a granite boulder for the middle of the floor, just to show that one can make flat stones from round ones, but life isn't long enough to do an entire porch like that.

Without a good supply of flat stones, I would be tempted to build the same porch right to the top of the perimeter wall, then pour a concrete floor inside the capstones. Concrete would rest easily on the sand and gravel fill, or even on a rubble fill.

If deep foundations are giving you pause, you could put a smaller stone porch on a "floating" concrete pad foundation. The concrete pad is at ground level, with good gravel drainage beneath it. Separate the porch from the house by a very narrow gap, enough to let the porch move while the house stays put. The gateposts (Chapter 2) and the barbeque (Chapter 4) were build on floating pads.

From the floor up, the porch is a carpentry job. I planned timber posts for this one. But you could add low stone pillars to hold the posts (as we did for Super-Hibachi), or omit the roof altogether and call it a deck. (Figure 8-6)

Photo 8-14 To finish off the joint, press down hard with a small wet trowel to get a smooth compact joint.

Figure 8-6 Adding pillars, a roof, and decorative planters should finish off your new stone porch.

Chapter 9

Planter

Richard and Rosemary Loeffler use their backyard the way Shriners use a convention hotel. With four children, two dogs, uncountable cats, around the clock basketball and a pool, a few little flowers hardly stood a chance.

The Loefflers moved here from New York City, so the backyard mayhem seems like peace and quiet to them; but Rosemary recognized that — size of the village notwithstanding — her flowers were suffering from urban blight. Some heavy protection was called for.

A planter was the obvious answer. Something stronger than a careening basketball, and higher than MacTavish the dog could raise his leg.

Stone was an easy choice. Grand Central Backyard is a stone's throw from Don Moodie's blacksmith shop, and from the site of an old mill. It was a busy place even before the Loefflers arrived. Much of that village industry had been built on stone foundations. And the stones were barely half-buried along the banks of the adjacent stream. It was there for the effort of picking it out of the beaver muck.

Design

There are an infinite number of variations on the basic planter shown here. It could be built into a surface like the planting beds in Chapter 6. It could be freestanding or tucked against a wall. Square or round. Terraced or flat, lower or higher.

With all that freedom, are there really any "proper" dimensions for a wall like this? Possibly not; but the builder should at least have some idea of what makes a wall fall down, and a rough rule of thumb to plan by.

A dry stone wall is usually only as sound as the bonding within it. A bond, in this case, might be a single stone that crosses the wall. If it reaches all the way from the face to the back of the wall, it ties the wall together at that point. If a shorter stone overlaps another — and the two together reach from the face to the back — then they too help bond the wall. The effective width of the wall is the width of these interior bonds. The wall could be five feet thick, but without bonding units, the wall is no stronger than the width of the face.

Soil stability, too, affects the soundness of the wall. If the soil is subject to flooding and drying, expanding roots, decomposing organic matter, then the wall must have a wide, flat base, and good interior bonding to withstand the movements.

Quality of the stone is also a factor. Large, flat stones, which overlap one another with a wide area of contact, are more stable than a small, round stone, which may touch the course beneath at only three points. (Photo 9-1) It's a simple matter of friction: the flatter the stones, and the heavier the stones, the harder it will be to pull them apart.

Consider, too, the external wear and tear on the wall, the use to which it will be put. A freestanding, decorative wall will be subject to little more stress than the weekly bump of a lawnmower wheel. A planter, on the other hand, must stand up to weeding and digging, and to the expansion of frozen soil within.

Finally, there is the relationship between the height of the wall, and an appropriate width. A tall, skinny wall falls over more easily than a short, fat wall. It has to do with the centre of gravity and the width of the base. A 5 foot wall on a 5 foot base is more stable than a 5 foot wall on a 1 foot base. You can improve the

Photo 9-1   *The side of the Lauffler's house was suffering from urban blight.*

odds by changing the shape of the wall. Curves and corners, for example, create a wider effective base. Tapering the wall, pyramid fashion, lowers the centre of gravity and improves stability.

Now the rule of thumb: To build a planter with reasonably flat stones, make the base as wide as the overall height. Make the top no wider than the best big stones. The best big stones, of course, go on top, and we want them to reach all the way from the face of the wall to the back. (Figure 9-1)

The Loeffler's wall was 18 inches wide at the base, and 18 inches high. In profile, it tapered from an 18 inch base to a 12 inch top, which was the widest span possible for the capstones. There were some larger stones, but they were not presentable enough for the top. We used them as bonding units, lower in the wall. And we used them in the base.

Had the rocks been more rounded, I would have widened the base to achieve the same height. Had the Loeffler's yard been less "active", I might have risked a slimmer wall, or smaller stones.

After all those variables, I should offer one constant, one elemental feature of planter design. Here it is: the planter was built dry — no mortar between the

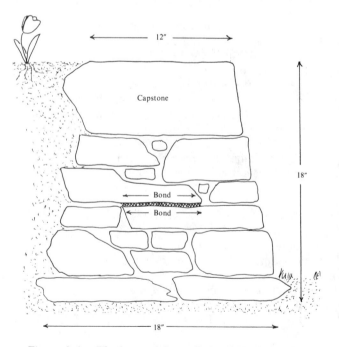

Figure 9-1   *The base of the wall should bed as wide as the height. Use the best stones for the cap.*

100

stones, no solid foundation. It's easier that way. And I believe it's better.

A planter must have dirt inside, and water. Otherwise, the plants don't grow. Damp dirt is fine in the summer. In the winter, it freezes. When it freezes, it expands. When it expands, it breaks the planter.

A mortarless wall, like this one, moves and flexes with the frost. In the spring, it settles back down to its original form, more or less. And there isn't any broken mortar to show that it moved.

If you must cement your planter's stones together, first build a solid foundation, Then place your left hand on this book, raise your right hand, and swear that you will faithfully clean out every bit of dirt in the fall. Every fall.

Preparation

This is one of those projects that takes longer to prepare than it does to build. The first job was to remove all the roots, sod and unwanted vegetation. There are two reasons. First, removing the dross by the roots saves a lot of tedious weeding later when this stuff worms its way to the surface again. Secondly, we want to start the wall below sod level. That is below the level where roots, moles, and decaying sod can shift the stones around.

Photo 9-2 Scrape out a shallow trench where the planter wall will be placed.

It took two Loeffler boys, a dull axe, and the neighbor's backhoe to grub out the tougher roots. But, by the time we finished, we were confident that there were no more magic beanstalks lurking under the surface. That was the hard part.

The easy part was scraping out a shallow trench where the planter wall would go. "Scrape" is the operative word. Not "dig". Digging loosens the soil. A heavy wall on loose soil would compress the dirt and sink a little. If it sank uniformly, it wouldn't matter. But soil harbours all sorts of perverse anomalies like rocks and hard spots. When part of the wall compresses the soil and part does not, the wall sags and leans. (Photo 9-2)

Remove the loose dirt, and scrape off the high spots to flatten the bottom of the trench. Don't fill in the low spots. The base of the wall should rest on undisturbed soil.

One of the more appealing aspects of building with stone (for some) is that a crooked wall looks as if it were planned that way. Stone can be the builder's version of blank verse, or abstract art. Skip the next step if you enjoy Andy Warhol movies. If, on the other hand, you like geometry, mowing the lawn, and straight stone walls, find two stakes and a string.

Drive in the stakes at the corners. Better still, drive them in a foot or so beyond the corner, leaving yourself room to work. Stretch the string from stake to stake. (Photo 9-3)

In the early stages, the height of the string doesn't matter. It is primarily a guide for horizontal alignment. Move it up the stakes as you progress. At the top, if you wish, you can level the string and use it for vertical alignment.

The Wall

The first course will be out of sight by the time the grass grows back, so start with the ugly ones. Broader stones will support more weight with less soil compression, so start with the large, ugly ones. Fill the trench with these big, flat stones. The wider the base, the better; and the grass will cover it anyway.

Raising the wall will be simpler if you tackle each course in this order:
1. Set the cornerstones
2. Fill in along the face
3. Fill in the back edge
4. Stuff rubble into the remaining gaps
5. Fill in the dirt behind
6. Put loose soil on top the wall, and brush it into the crevices.

Start with a rough sorting of the stones at hand. No need to move them around. Just have a good look.

Photo 9-3   Use stakes and string to help keep the wall laterally and horizontally aligned.

Take special interest in their thicknesses — the heights of the stones when laid on the flat.

Each course, or layer, of stones in the wall should have a roughly level top. This is rarely possible along the entire wall, but grouping similar heights has an advantage. A flat top in this course makes a flat base for the next course. One larger stone in the middle of a course creates two steps that somehow have to be bridged in the succeeding course.

You can change levels in the middle of a course. But it takes a tapered rock, or a small filler, to do so without running a continuous vertical joint. The fastest way to build is with uniform courses. (Figure 9-2)

The strength of the wall, the thing that ties it all together, is the overlapping of stones. This is a first principle in any stone wall. The trick is to span the joints in the course below, staggering the vertical joints. (Photo 9-4)

Start at the corner with a squarish rock. Then find some face stones of the same height. Put one next to the corner stone. If it ends near a crack in the course below, discard it and try another, longer or shorter than the first one.

When the course is complete at the face, repeat the process along the back. The back doesn't have to be as neat and presentable as the face, but it should be as strong as the face. And, for the same reason that you tried to make an even top along the face, build up the back to the same height as the face. If the top of the wall slants forward or back, the next course will be less stable, and much harder to fit.

Now you'll have smaller holes and gaps between the face and the back. Stuff any sort of rubble in these holes. But, again, don't let the centre stick up higher than the face or the next course will have a hump to cross. A hump in the centre will make the next course less stable, and harder to fit.

Figure 9-2   Try to keep each course level. If you have an uneven stone or step (left), you can bridge it with a tapered stone (center), or use small fillers to make the next course level (right).

Photo 9-4   Strength in any stone wall is achieved by overlapping stones providing a bonding effect along the entire wall.

As you work along the face, align the front of the face stones with the string. Remember to set each course back slightly to allow for the taper. (Photo 9-5)

Also at the face, remember to include a bonding unit somewhere in every course. Distribute these along the wall, and from bottom to top. If you have a limited number of large stones, save them for the higher courses. The nearer the top, the more important it is to tie the wall front to back. Indeed, the most important bonding units in the wall are the capstones.

If none of the stones are broad enough to span the entire width of the wall, use two overlapping stones. Set one at the face, reaching back across the centre of the wall. On the next course, set another large rock at the back, reaching far enough forward to overlap the face stone below.

Only by the sheerest luck will any stone sit solidly on the wall at first fitting. The odds are better that a toddler will get his boots on the right feet. Most of the stones that you put on the wall will wobble. The solution is a handy pile of small, flat stones used as wedges or shims. Break up a larger rock if the soil doesn't contribute enough shims. Or give the nearest toddler a bucket and tell him what you want. Kids know where all the best rocks are.

Set a stone on the wall where you think it ought to be. Then wobble it. The wobble means that the stone is

Photo 9-5   As you work along the face, align the front of the face stones with the string.

103

resting on two points and is trying to decide on a third. Or, it means that the stone is supported near its centre, and pressure around the edges upsets the balance. A wobbly stone needs at least one more point of support near its perimeter.

The builder has a choice. You can tip the stone back and put a shim under the front edge, or tip it forward and shim from the back. From the back is better. As the wall moves with frost or weeding or whatever, the shims can wiggle loose. A shim at the face of the wall might fall out. A shim at the back, inside the wall, has no place to go. It can't fall out. (Photo 9-6)

Photo 9-6 Use shims along the back side of the wall to keep the stones level.

Wobble the stone forward. In other words, put your weight on the front edge of the stone. The back edge will lift a little. Jam a shim in the gap until it's tight. Try again to wobble the stone. It may need a shim under another corner. If the gap is too high for the shim, don't push the shim in too far. Shimming a stone under its middle is no help at all. Look for a fatter shim instead.

As each course is completed, add more soil to the planter. This is not to pre-empt the gardener, but to help hold the backside of the wall in place. The soil sifts into the cracks and firms up the fit almost as well as the shims. Put more soil on the top of the course, and brush it into the cracks. Eventually, it will find its way into the gaps and help stop the wobblers.

Capstones

As the wall nears final height, level the string. You can buy a line level, made to hang on a string like this. Or, you can align the string with the bricks in the wall, the siding, or the neighbor's fence.

Now the object is to align the top of the capstone course with the string. The temptation is to set the heavy capstones first, then level the top by adding small fillers above them. Don't. Measure the thickness of the capstones. Measure the gap between the string and the top of the last course. Then put the filler stones on the wall, *under* the capstones.

Stone walls fall apart from the top. The heavier the stone, the harder it is to dislodge. A heavy capstone holds the little rocks under it in place. A little rock, left on top, would fall off and break the lawnmower. Adjust for the final height by filling in *before* the capstone course.

The capstones are the most vulnerable part of the wall. All the sitting, climbing, weeding and digging take place there. An individual stone can be loosened with all the activity, but there is only one way it can move. It can't go back into the planter — the soil blocks it. It can't move to either side — there are more heavy capstones on either side. Gravity won't let it rise. The only choice it has is to stay put, or wiggle forward until it falls.

For that reason, and to bond the wall better, set the capstones *across* the top of the wall, not lined up along the edge. The farther back you set the stone's centre of gravity, the less likely it is to fall off. (Photo 9-7)

Photo 9-7 For greater stability of the capstone, set them with the longest side across the wall, not parallel to the edge.

Photo 9-8   You'll be shoveling in your top soil in no time, and getting ready to plant your favorite flowers and shrubs.

The truth is, it took longer to type these few pages than it did to build the planter. Two hours after uprooting the stump, Richard was raking in the final layer of fill, and MacTavish was applying his stain of approval at the corners. (Photo 9-8)

It's fast because there is no mortar, the foundation is the earth, and fitting can be accomplished with shims instead of chisels. But it will last as long as any mortared planter. Make it any shape you like. Just remember the principles and the rest you can make up as you go along. MacTavish would approve.

Chapter 10

Mudroom

If the world ever beats a path to our door, the first thing they'll do is track in mud. You can seal the doors, or surrender and plant the kitchen floor with rice, or build a mudroom.

The problem is what to put on the floor. Wood can't stand the water. Carpets are a washout. And a gang of kids on ice skates would cut vinyl to ribbons on the first cold Saturday morning. Concrete would work, but it's not very dressy for an entrance. Stone answers all those objections.

Design

That was the good news. The bad news is that stone floors can be heavy, and cold. The design has to overcome heavy and cold.

The weight problem is not so hard to solve as it sounds. If you already have an enclosed back entrance, chances are that it was made by covering an outside porch. If you have to build a new entrance area, chances are that you'll find it cheaper and easier to add on to the outside of the house rather than subtract space from the interior. Either way, if the mudroom is outside the basement wall, you can easily include a stone floor built on compact fill.

This won't be practical above a basement. But for an exterior addition, an enclosed porch, or over a crawl space, it's ideal.

Winter is the other problem. Outside the basement walls, the earth is cold. If the floor is built on an earthen base, then it too will be cold. That's the way it used to be. Then someone (probably someone who pads around the house in stockinged feet) invented rigid foam insulation.

I had some initial doubts about the ability of this stuff to hold a heavy floor as well as the human traffic. I knew that engineers were building highways on it, but what would a highway engineer know about a scrum of ten-year-old hockey players. I am now assured. In the past eight years we've had everything but the pony across this floor: wheelbarrows, bikes, skates, dogs, stilts, ten goats and a lamb. And there is not a single crack. After the weekly scrubbing, it looks as good as the day it was done.

The secret — if there is one in this simple design — is thorough compaction of the soil/sand base. The fussy part is not the fitting of stones, but the leveling.

Materials

Flat stones
Sand fill
Rigid foam insulation
Mortar mix

The ideal stone would have at least one flat face, and lots of straight edges. One flat face is not so hard to find. Limestone and sandstone naturally split along straight planes. If these are badly weathered, the faces may be pitted and rounded; but don't stop looking at ground level. Dig a little deeper. The same stone, protected by a thin layer of soil, may have just the flat face we're looking for.

Even boulders, even granite boulders can be cut into floor stones. Look for the grain — straight lines that run around the stone like latitude lines on a globe. Pick a prominent line near the centre (the equator). Set the chisel parallel to this line and hammer straight along the plane. Pace yourself . . . results will not be hurried. And don't go for the heavy weapons; tougher splitting calls for smaller chisels (the narrower cutting edge concentrates the force on a smaller spot). It is important to keep the chisel at a consistent angle, in order to send the fractures along a single plane. Eventually, you will see the grain widen into a crack, or you will hear a lower note in the ring of the chisel. Now a few hard blows will finish the split.

If finding, or making flat stones is a pain, take heart from the fact that you don't need many. This 8 foot by 5 foot room takes 40 square feet of faces. That's about one pickup load of stones.

The sand is for fill. It needn't be fine, masonry sand. Some gravel in the fill is acceptable. Do avoid fill that contains any loam or clay, however. The quantity of fill required depends on the difference between ground level and floor level. Add an extra 6 inches for any sod to be removed, and calculate the total volume.

Foam is sold in sheets 8 feet long, and up to 4 feet wide. I used a 2 inch thickness (2 inches of blue foam is rated R 10 for insulation value). For quantities, calculate the area of the floor, then add enough to cover the sides of the filled area, from the floor to the ground.

Preparation

Start with a roof and walls. If you're building from scratch, leave out the floor. If you're starting with an existing room, remove the old floor and check the wall construction.

From ground level to floor, the wall, in fact, will be a retaining wall. It must hold back the outward pressure of the fill. Regardless of how the upper walls are built, that lower portion of wall must be solid, damp-proof, and rot resistant.

An existing block or concrete foundation would be ideal. But do keep in mind that the floor level can be no higher than the foundation. If you want to raise the floor to match interior floor levels, it may be necessary to augment an existing foundation with an extra course or two of blocks.

If the mudroom is an entirely new addition, plan to bring the foundation up to the desired floor level.

Let's begin, then, with a closed-in room, and a solid perimeter wall. There's an earth-bottomed hole where the floor should be. The first step is to compact the fill already in place. If it's a new addition, there may be sod down there. Remove it. Remove any loose topsoil or organic material. Now tamp.

You can rent a power tamper, but for a small space like this it may be easier to use a length of 4 × 4 timber. Even a small log will do. All it needs is a small, flat end, and some heft. Stand upright, hold the tamper in two hands, and tamp. Cover the whole area in a regular pattern, overlapping the blows. This existing fill may already be compact. In which case, you won't make much impression. But you may find that the earth will be softer around the edges, where the foundation excavation was filled. Give extra attention to the edges.

Now add insulation to the inside of the foundation walls. Measure the distance from the top of the fill to the desired floor level. Cut the sheets of foam to fit this area. The foam cuts easily with a kitchen knife.

It's tempting to extend the insulation lower down the foundation wall, but if this means disturbing compact fill, forget it. The fill is more important. You can insulate the foundation wall below grade by excavating outside the wall and applying special foam sheets to the exterior. Ask your building supplier for advice on materials and installation.

The cut edges of foam crumble easily. Turn the cut edges down, and leave the smooth, factory edge up at floor level. There is a special foam adhesive which you can use to glue the sheets to the wall. Or, you may simply hold the sheet against the wall and add enough fill at the bottom to hold it in place temporarily. If you do it without the adhesive, be extra careful to keep the fill from trickling in behind the sheets. We want them flat against the wall, and tightly butted, end-to-end.

The first layer of new fill is the leveler. Shovel in enough sandy fill to cover the bottom of the hole. A layer more than 3 or 4 inches deep will be harder to compact. So, if the bottom is very uneven, level it with several separate layers of fill.

Rake this first layer as flat and smooth as you can make it with the eye. For more accurate leveling, cut a straight-edged board slightly shorter than the width of the hole. Set the board on edge, across one end of the fill. Then "saw" the board back and forth, moving it forward each time. The high spots will be ploughed up ahead of the board, filling the low spots as you pass over them. Work the board from end to end, side to side, and corner to corner.

Now soak the layer of fill with the hose, washing the sand down into every hole. Tamp the wet surface hard.

The soaking and tamping will rough up the surface. Let it dry a little, then level it again. Add another layer of fill if the surface is not yet flat. Repeat the leveling and compaction routine.

When you are satisfied that the surface is as hard and flat as you can make it, cut the foam to cover the surface. Start with the full-sized sheets, butting them snugly into the corners and against the already insulated walls. Lay each sheet at least twice. Press it down firmly the first time, feeling for the soft spots that suggest there might be a hollow beneath. Then raise the sheet to see what impression it made on the damp sand. Now you can see the dips and hollows. Sprinkle more sand in the low spots and try the sheet again.

The purpose of all this fastidious leveling is to eliminate any hollow spots under the foam. The foam can bridge the hollow now, but when the rest of the fill and the floor have been added, the extra weight may bow the foam into the hollow and allow subsidence above.

With the surface entirely covered by insulation, add the rest of the fill. Again, add it in thin layers, wetting and compacting each layer. The difference now is that you don't have to level each layer so meticulously. A rough rake-off is enough.

The first layer of fill atop the foam should be added carefully. Adding weight on one part of the sheet can raise an edge someplace else. If the sand gets under an edge, the sheet may ride up and leave new hollows beneath. Put the first few shovelsful on the joints, watching to be certain that no edge rides up. Once the edges are covered, the rest of the fill can be dumped in any old way.

Continue filling in layers until the surface is a stone's thickness lower than the desired floor level.

Floor

The first requisite is a level edge, a reference point from which the stones can be leveled. The best, and most obvious answer is the top edge of the foam sheets around the walls. If this edge has somehow been damaged, or is obscured by the upper wall, mark a level perimeter with a chalk line, and nail a temporary strip of wood below the chalk line.

A straight board, long enough to span the room, will rest on the top edge of the foam (or on the temporary strip) at both sides of the room. (Photo 10-1) It's also handy to have a shorter board, one that fits more easily into the corners. The bottom of the board defines the plane of the floor. In other words, the tops of the stones have to be aligned with the bottom of the board.

Start in the corners. If you're a novice stonecutter, the corners may be the most difficult part of the fitting. Look for a stone that has two edges forming a right

Photo 10-1   A straight board, resting across the floor, will provide a base for leveling the stones.

angle. There aren't many — Mother Nature is not that obsessed with 90 degrees. Here's how to cut them.

If possible, pick a stone that already has one straight edge. Place a rafter square on the surface and mark the cut at right angles to the existing edge. Lay the stone flat on the sand. Then score the mark with a wide chisel. I prefer the 3½ inch brick chisel for this job. It's easier to keep the wide blade running straight along the mark. Hammer gently — the thinner the stone, the softer the touch. Work back and forth along the line until the chisel wears a groove in the stone surface. Now you can hammer a little harder. Not all in one spot, though — continue to move up and down the line. In time, a crack will appear. If you've been patient, and lucky, the crack will run along the bottom of the groove. A few more whacks of the hammer will take off the excess.

The corners are first because those are the ones you're most likely to have to cut. The next priority is size. There is a reason for this order. The biggest stones should be placed in the most heavily traveled areas: in front of the doors, and on the path between the doors. Large stones are more stable than small ones. Their weight keeps them from tipping under uneven pressure. And the larger surface distributes weight over a wider area, making a big rock less likely to subside than a small rock. So the best way to ensure that the biggies fit where you want them is to put them in first. Then fit the smaller rocks around the big ones.

Level each stone as you lay it. The reference, remember, is the board, resting on the perimeter of foam. Set the board over the stone and judge how far the stone must be raised or lowered to meet the underside of the board. Tip the stone on edge and rearrange the sand beneath. The stone might well have an uneven bottom. Dig holes for the bumps and fill in the rest.

Wet the sand and reset the stone. Chances are it's now closer to the mark but still in need of adjustment. Raise it again. You can see the impression made by the stone on the bed of wet sand. The smooth areas have been flattened by the stone. The rough areas are the hollows, where the stone was unsupported. Dig and fill accordingly. A small rake, or a trowel, make ideal diggers and fillers.

Reset the stone. Stomp on it to compact the sand beneath it. If one edge is high, you may be able to stomp it down to the desired level. If the whole stone is high, even after stomping, you'll have to raise it again and rake more sand away.

Raking sand is like shaking the catsup bottle. The stone stays aggravatingly high, until you flick away that last grain of sand. Then, somehow, the stone drops half an inch below the mark. Glub!

The inevitable last step is to raise an edge that last half inch. You're done with tipping, digging and filling

when you get this close. Get a crowbar, another shovelful of sand, and the garden hose.

Put the crowbar under the low edge of stone and raise it. Raise it until it meets the marker board exactly. That leaves a gap under the stone. Dump the sand there and wash it under the rock with a hard blast from the hose. Wait a moment — to let the sand settle — then ease off the crowbar. The stone may drop a fraction. Raise it again with the bar, dump more sand, and flood it into the gap. (Photo 10-2)

When the stone is level under the board, turn the board in another direction and check the stone surface again. You may have to do more adjusting with the bar, the sand and the hose, until the stone matches the board from several angles.

Photo 10-2   To raise a stone you will need to lift an edge with a crowbar, fill the gap in with sand and wash it down in with a hose.

Fitting the Mosaic

Fitting stones edge-to-edge is a visual skill more than a manual one. To some, it comes naturally. They can look at the zig-zag edge of a half-built mosaic, then pick out the stones that will match the angles. The rest, who will try to fit square rocks into round holes every time, must do a lot more cutting.

Imagine an edge, formed by the straight edges of two rocks already in place. The two edges make an angle. The stone that will fit in that space has two sides with matching angles.

That's the first criterion. The second is the lengths of the two sides that form the angle. If you can manage to match the lengths of the sides, as well as the angles, then the next row will be that much easier. (Figure 10-1)

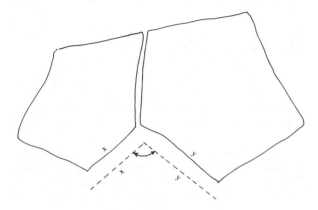

Figure 10-1 An interior floor should be fitted together as neatly as possible. This will require fitting the stones by matching the angles, and if possible the lengths of the sides.

A close fitting will require some cutting, no matter how keen your eye might be for shapes. Lay the rock that almost fits in the space intended. Because of the angle of the sides, or some protrusion, the rock doesn't fit snugly against its neighbors. Use a straight edge and chalk to mark the cut, parallel to the edge of the rock already in place. Then move the stone back onto the sand and cut it just as you cut the corner stones. Again, use the wide chisel to score a groove. (Photo 10-3)

A common barrier to the best fit is a rounded edge. If the edge doesn't drop away squarely from the face, the two rocks may meet below the floor surface. Though the fit may be tight, the surfaces of the two adjoining stones would be separated by a wide mortar joint. It looks like a poor fit even if it isn't.

Square off rounded edges, or undercut them, by chipping away at the protrusion with the narrow chisel. Start the chisel below the face, and angle it back into the rock. The first few blows will be needed to make a niche for the chisel, a ledge to keep it from skipping over the bump. Once the chisel has a foothold, hammer away. (Figure 10-2)

Fit the edges, then level the stone. We started in the corners, and then placed the largest stones in the heavily traveled area. Now work around the larger stones, filling in the most visible areas next. In a mosaic like this, the hardest stone to fit will be the last one. So far, we've been fitting each new stone along an existing

edge. The shape of the new stone's backside doesn't matter. The last stone, however, has to fit the hole around all its edges. The best strategy is to anticipate this difficulty and do the visible areas now, and leave the last gap somewhere out of sight. In our case, we planned to end the fitting under the bench. If necessary, we could fill the last hole with little pieces, knowing it would never be walked on, and knowing it would be hidden under a pile of muddy boots for 360 days of the year.

If your floor doesn't happen to have a dark corner, you can cut the last stone to fit the space, or re-shape the space to fit the stone. (See Chapter 1)

When the entire floor has been fitted and leveled, flood it once more to wash the sand into any remaining gaps. Then put it to use right away. If your household is anything like this one, there has been a constant parade over the work since it began. But — never mind — now they can use it officially. The last step, the mortar between the joints, can be postponed for a week or more.

Photo 10-3 Use a wide chisel to cut off proturusions on the stone which prevent a good fit.

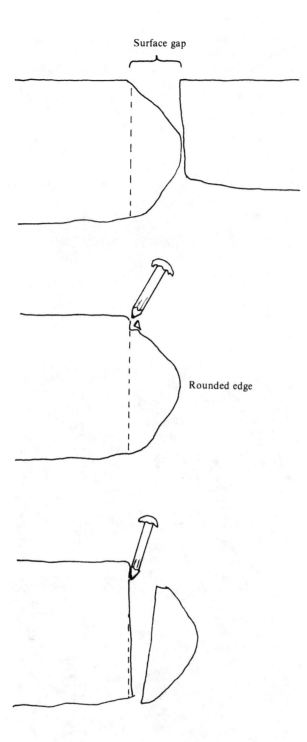

Surface gap

Rounded edge

Figure 10-2 A stone with a rounded edge will not butt tightly up against the next rock, leaving a large gap (top). To chip off the rounded edge, chisel it off with the chisel angled toward the stone (center), and the chip should come away fairly vertical (bottom).

The week's interlude is to allow any early subsidence to show itself, to find all the toe-high edges, and to trample any poorly-seated stones into conformity.

At the end of the week, go over the entire floor again with the marker board and the level. You may also need the sand and water treatment in spots, raising stones that have been trampled down too far. Now is the time to barricade the door and banish the cat to the basement. You don't want anybody walking on the floor for awhile.

When the surface is completely level, wet all the joints with the hose, and stir up a batch of mortar. You can make the mortar from bagged mix, or mix your own, but do make the batches small. This is slow, fussy work, and the consistency of the mortar must be right on — not too wet and not too dry. If it's too wet, it will wash onto the surfaces and discolour them. If it's too dry, it won't flow into all the crevices deep in the joints. The ideal mix is the consistency of soft butter. You should be able to slice through a 3 inch pile of the stuff and leave a clean, vertical face behind — no slump from the wet, and no crumbling from being too dry.

If you make your own mix, use screened sand. Pit run sand may contain stones too big for the narrow joints. The mortar won't work deep into the cracks if it's hung up on bits of gravel.

Filling joints is best done with two large trowels. Scoop up half a load of mortar on one and place the edge of the trowel at the crack. Use the other trowel to slice off thin slabs of mortar, dropping them into the gap. The two trowels form a "V", with the mortar between them. The trowels keep the mortar off the surfaces of the stones, funneling the mix between them into the narrow gap.

When you've dumped the load, use one of the trowels to work the mortar as deeply into the crack as it will go. Slide the trowel up and down in the joint, chopping the mortar into the crevices. Wet the tool frequently, so it doesn't drag mortar back out of the joint. It should slide in and out easily.

Continue filling the joint until the mortar stands about ¼ inch higher than the surface of the stones. Don't try to smooth it. Just pack it in and leave the top surface chopped and rough.

The small stones that fill odd corners between the big ones are a special case. Lift out the stone. Remove a handful of sand. Fill the hole half full of mortar. Now squish the stone into the mortar. Push it down until its surface is flush with the floor. The mortar will ooze up around the edges of the stone. If you judge the volume right, it will ooze up and fill the entire joint. If not, top it up like any other joint, with the two trowels working in a "V".

Indoors, against damp stones, the mortar should

not begin to stiffen for at least an hour. That's enough time to use the whole batch of mortar. Perhaps enough to mortar the entire floor. Check the first joints periodically. They're ready to finish when the mortar offers some resistance to finger pressure. If you trowel the joints too soon, the mortar will ooze a watery drool across the stone, spoiling the appearance.

A full-sized trowel is too wide to finish joints like these. You need a flat blade, about 1 inch wide. You can buy one ready-made. But I've used a putty knife and a butter knife with equally good results. The one in these photos is a scraper that was on sale in the paint department.

Draw the blade along the ridge of mortar, pressing down hard to force mortar farther into the joint, and to bring a sheen of water to the surface. (Photo 10-4) You don't have to smooth all the mortar — just a seam down the centre. Flatten it out until it's flush with the stone surface. Let the little ridge of excess mortar collect at the side.

Wet the trowel when it starts to drag, and go over the seams until they're slick and flat.

When all the joints have been finished, go back and remove the crumbly ridges at the sides of the joints. Use the edge of the trowel to slice the excess away from the seams. Carefully! You don't want to mar the finished parts, or drag mortar out of the smooth surfaces.

Photo 10-5   We finsihed the mudroom off with oak panelling which covered the top edge of the foam insulation.

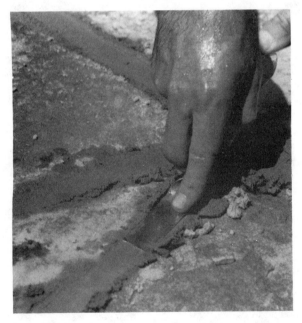

Photo 10-4   To make smooth gaps use a small wet trowel and press down the mortar as you move along the joint.

Cut the excess away. Then gently push it aside, on to the flat rock surface. By now, the ridge of excess should be dry and crumbly enough that it won't stick to the stones. Don't try to clean it up now, though. Leave it alone for several hours. Then, when the pointed joints have hardened a bit, you can brush away the crumbs with a soft broom.

After all the waste has been removed, wet the surface with a fine spray, and cover it to keep it damp for at least three days. A sheet of plastic will keep the moisture in, or you can use damp burlap bags.

For those three days, keep the barricades up and the heavy traffic away. The cat can't harm it now, but a heavy step in the wrong place may still tilt a stone and break the mortar joint. If you absolutely *must* cross the surface, step gently onto the centres of the large stones only. Stay away from the joints and away from the smaller stones.

Alternatives

We finished this mudroom with an oak panelled wainscot, boot bench, and coat rack. The bench rests on antique laundry tub legs, and the wood is all scraps of hardwood flooring. (Photo 10-5)

The panelling is handsome, cheap and durable. More to the point, it brings the wall out just far enough to cover the top edge of foam. The end of each vertical panel was cut to conform to the top of the stone.

The alternative is to use the more common gypsum board panels, or drywall, with a baseboard and quarter-round moulding. (Figure 10-3) Half inch drywall, plus ¾ inch baseboard and ¾ inch moulding would cover the 2 inch foam. The major drawback is

the difficulty of carving the quarter-round to fit flush against the stones. Given the choice, I would recommend wainscot.

The one improvement I might attempt, if I were doing the job again, would be a recessed well for the doormat. A one inch recess would hold the mat in place and keep it out from under the door.

An alternative I would not recommend is leaving the joints unmortared. This works fine on the outdoor patio. But indoors, there would be a constant litter of sand from the joints. The floor would be almost as solid, but housekeeping would become more difficult. We have an unmortared flagstone floor in another small room, but the mortar makes a cleaner job.

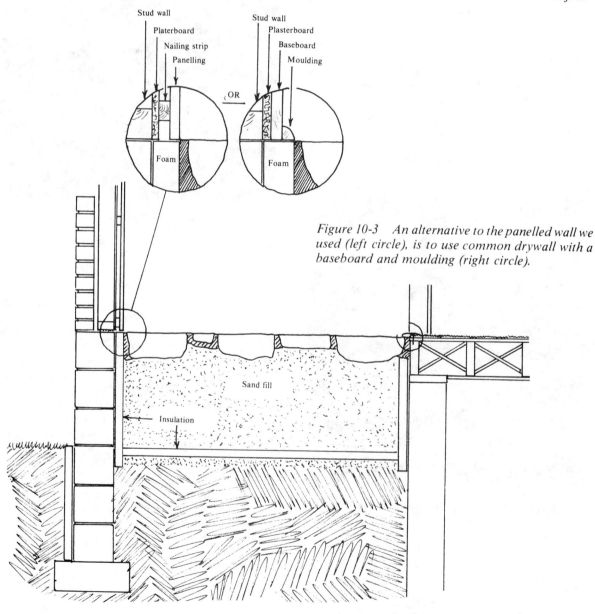

Figure 10-3   An alternative to the panelled wall we used (left circle), is to use common drywall with a baseboard and moulding (right circle).

Author's Note: In some lakes, environmental agencies regulate any work done at or below the waterline. Your local building inspector will know what permits might be required.

Chapter 11

Dock

"Tommy" Thomson retired with the same enthusiasum that most of us reserve for root canal work, or income tax. He would really rather be working; which is why he has a hand-made railroad in the back yard, a basement full of contraptions that have to be tinkered with, and a monumental itch to be doing.

There is no way that a steam-driven personality like this one could look at the shore of his lakeside home and not think of one or two ways to improve on mother nature's mishmash. The lawn dropped away to meet the lake at an abrupt rock ledge. There was a tatter of grass at the top, and a jumble of fallen rock below. The water, rising and falling as much as 30 inches over the season, alternately covered and exposed the loose rock at the bottom of the ledge. Tommy, a boater, and Marion, a swimmer, agreed that access to the water could be better. The concrete ramp and the waterlogged raft just wouldn't do.

And, if looking at the lakeshore wasn't motivation enough, Tommy only had to turn to the right to see neighbor Bob Edgington's fine stone dock.

Design

In the northern lakes, nature imposes some strict conditions on the building of docks — conditions which would make a better dock in any climate, but which are essential where ice and spring currents can sweep away a flimsy dock like last fall's leaves.

Most of the damage is done by spring ice, moving along the shore. So the safest dock is one that follows the natural contour of the shoreline, allowing the heavy ice to slide past unimpeded. The face of the dock should be relatively smooth, rising from its base below the waterline in a gentle slope that leans back towards the shore, leaving no projections to catch the ice.

Fortunately, the Thomson waterfront was deep enough that a shore-hugging dock was feasible. Five feet out from the natural ledge, the water was deep enough to float anything on the lake in summer, and there would still be plenty of water for Tommy's pontoon boat in the fall, when water levels dropped. (Photo 11-1)

The front of the dock would follow the curve of the ledge, starting from the existing launch ramp at one end. The deck, or top of the dock, had to be about 8 inches higher than the highest water level. This would still leave the deck 24 inches lower than the lawn. The difference in height necessitated a low retaining wall at the back of the dock, and three steps up to the lawn.

The retaining wall and the steps were designed to incorporate parts of the natural ledge. (Figure 11-1) Part of this was laziness — every rock that nature had laid was one less for us to carry — but the big advantages came down to strength and appearance. The wall was stronger for resting on a bedrock foundation. And the marriage of new masonry with nature's own would give the design a more natural look, a minimum of intrusion.

Much of the planning centered on the rising and falling of the water. We had to leave enough depth to float a boat in the dry times, and raise the deck high enough to keep Tommy's feet dry when the lake was brim-full. Moreover, we wanted to include a series of dockside steps so the boat could be boarded easily, regardless of the water level.

The starting point was so find the high water mark, and use that as a base from which all other measures could be taken. Doug Clark, neighborhood jack-of-

Photo 11-1 *The water depth at the edge of the dock was easily deep enough for mooring a boat.*

all-building-trades, was also the informal foreman of the Thomson dock. Tommy wasn't sure where his own high water mark was, but Doug had a dock just down the shore and Doug knew exactly how high the water rose in spring. So we measured the height of Doug's high water line above the current level, then hustled back to Tommy's to mark the ledge.

Now we had an annual high water mark on the ledge. We added 8 inches to that, to keep the floor higher than the average wavelet, and marked a line for the top surface of the finished dock — the floor level. Nails, driven into the cracks of the ledge, marked the level more permanently than the usual chalk and gave us something to hook the mason's line to. And, for once, it wasn't necessary to level the line. Measuring up the same distance from the water at either end ensured a level, so long as the lake didn't tilt.

So much for the top. What about the steps? The starting point is to measure the boat. The deck of Tommy's boat rides 12 inches higher than the water. A comfortable step is 8 inches high. Eight plus twelve makes twenty. When the water is at its lowest level, the captain should be able to step *up* 8 inches to the lowest step. So, the lowest step should be 20 inches above the low water mark.

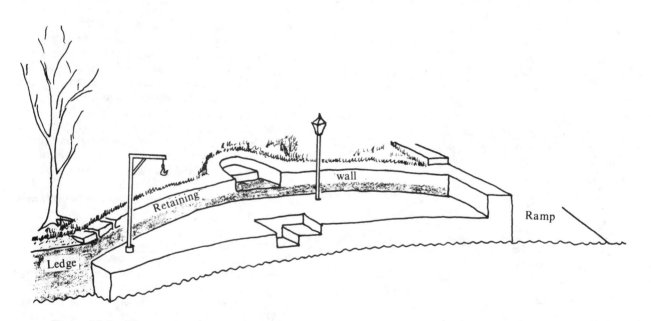

Figure 11-1 *The retaining wall and steps of the dock were designed to incorporate parts of the natural ledge.*

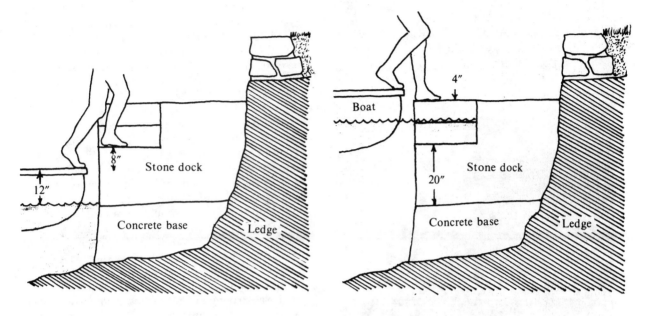

Figure 11-2   The dock steps had to be designed to accomodate both high and low water, which has an 8 inch step up at low water level (left), and a 4 inch step down at high water (right).

The dock, remember, is to be 8 inches higher than the high water mark. So, at high water, boaters step *down* 4 inches to the dock. In reality, for most of the summer, the deck and the dock are at the same level. (Figure 11-2)

In general, the low water level determines the height of the bottom step, and the high water level determines the height of the dock's surface. Between those two levels, the steps are set at 8 inches intervals.

Changing water levels imposes one more problem of design: the risk of trapping water inside the dock. The stone wall is not impervious. One way or another, water will find its way behind the face. The lake will leak in, or ground water will fill the porous space behind the stone. Usually, this back water will remain at about the same level as the lake, and no problems result. The problems come when the lake level changes — leaving tons of water trapped behind the face. Unless you're prepared to build the thing as strong as a dam, the uneven water pressure will damage the masonry. If water trapped behind the face is then allowed to freeze, damage is assured. The trick is to let the trapped water escape as the lake level drops. The trick, in other words, is drainage. We planned a series of drains through the masonry at the low water line.

The rest of the design was as much for fun as for function. Tommy insisted that the derrick was needed to life the motor off the boat, though the Edingtons pointed out that it looked a lot like theirs. The light was Tommy's idea. The timber bumpers came from Doug's bent for carpentry.

Finally, we planned to build when the water was at its lowest level. This way, the concrete base which was to be poured to the waterline, would never be seen. And the amount and cost of the concrete would be minimized. Unfortunately, the water in this particular lake is at its lowest level in the autumn. We set to work in November, up to our waists in water that seemed to be about one degree shy of skatable.

Materials

Concrete forms (or plywood and lumber to build them)
Wire and cutters
Scrap lumber and common nails
Reinforcing steel: rods, pipe, fence posts, etc.
Concrete
Plastic pipe
Pipe casing
Stone
Cement: portland and masonry
Masonry sand
Backfill

And, of course, you will need the usual array of trowels, chisels, hammers and buckets. For a job this size, an electric mixer will be worth borrowing or renting. Heavy tools, sledge hammer, wedges, and a pry bar, are essential for work on bedrock ledges. And a raft, or a floating dock, will be handy for holding stone and for standing room while you work on the outer face. (Photo 11-2)

Photo 11-2 A raft or floating dock is handy for working on the outer face of the dock.

A prefabricated concrete form is usually made up of modular sections. Each section is 2 feet wide and 4 feet, 6 feet, or 8 feet long. The forms are wired or bolted together, side-by-side or end-to-end, to contain the wet concrete until it has hardened. Then they can be removed and re-used. You can make your own forming sections from heavy plywood on 2 × 4 frames, but this can be expensive if you're only going to use them once. The alternative is to rent form section from a small contractor or a rental firm. You'll need them for about three or four days. The number depends on the size of the dock and the depth of the water. In our case, the full form had to be 36 feet long and 2 feet high at the deepest point. In other words, 4½ sections long, and one section high.

The wire and cutters are for joining the sections together. The easiest and cheapest is "fence" wire, or "black" wire, obtainable from farm supply stores.

The scrap lumber is to help hold the forms in place. Not only will the concrete want to push the forms out of line, but wooden forms have a nasty habit of floating away as soon as your back is turned. We'll use the extra lumber to make some temporary "feet" for the forms. Rocks, piled on the feet, sink the form to the bottom and hold it there. Any cheap lumber will do, but 1 × 4 or 1 × 6 are the handiest sizes.

Steel reinforcing ties the concrete base to the bedrock ledge, and anchors everything to the lake bottom. It's unlikely that seven cubic yards of concrete would float away, but ice and erosion are constantly rearranging the shore, splitting huge slabs of bedrock away from the ledges. Without strong ties, there s a greater risk that the dock, too, could be moved. We used bar stock, fence posts, rebar steel, anything that Tommy could extricate from the delightful pile of exotic junk behind his shop. I can't tell you how much steel we used. I was standing in the lake at the time. Icy water was pouring over the top of the chest waders and the wind was driving an early snowstorm into my face. I lost count. It seemed like 10,000. It was probably about a dozen rods, each 4 feet to 6 feet long.

The concrete base goes underwater. You pour it into the forms, displacing the water. There are a few tricks to pouring concrete underwater. The first is to pour quickly, displacing the water before it has time to dilute the soft mixture and wash it away. More about that later. For now, accept my word for it that the common or garden backyard mixer can't make concrete fast enough to fill underwater forms. For this job, call in a ready-mix truck.

How much to order? The rougher the bottom, the tougher it is to estimate the volume accurately. Estimate the *average* depth of water inside the forms. Length × width × average depth (measured in yards) will give you an idea of about many cubic yards of concrete will be needed for the base. Order some extra, just to be sure; or find a concrete company that offers on-site mixing and metered volumes (you pay only for what is used). Doug and I neglected to order extra and nearly turned his half ton truck into a four ton sculpture. We underestimated the pour, and the ready mix truck wasn't available for a second trip. So, Doug drove 14 miles, filled the back of his pick-up with wet concrete, and crept back to Tommy's with creaking springs. Too fast and the concrete slopped over the sides of the truck — too slow and nothing slopped . . . it just sat there getting harder. Order extra. It's so much easier than unloading a truck with a chisel.

Plastic pipe forms the drains that allow trapped water to flow through to the lake. We used 2 inch pipe at the low water line. First, because there was a lot of it lying around. And, secondly, because we feared that anything smaller might get plugged with silt or weeds. Smaller, plastic hose formed the above-water drains through the retaining wall. We cut the 2 inch pipe into 3 feet lengths and spaced them about 6 feet apart.

Pipe casing, in larger diameters, is a handy way to house the built-in extras. We used a length of 4 inch perforated, flexible drain pipe to carry Tommy's water supply lines under the dock and out into the lake. This 4 inch pipe (sold at building supply stores) is large enough that the smaller water lines can be pulled through for maintenance or replacement. A piece of 7 inch well casing formed the base for the derrick. Any heavy pipe would do the job, but it must be long enough to reach from the deck right down to the base, especially if the derrick is meant to carry any weight.

Quantities of stone, sand and cement depend on the size of the dock. Have a whole truckload of sand delivered. Haulage is usually the most expensive part, so you can't save much by buying smaller quantities. And any leftover sand can be used for backfill or beach.

Buy cement in small quantities, twice as much masonry as portland. Keep it dry and well covered.

The stone, in Tommy's case, was readily at hand. We dug up a lot of usable slabs while leveling the bottom for the concrete. The rest came from the ledge itself, and from a nearby shoal. As usual, we hauled twice as much as we thought we might need, and then went back for more to finish the job. We weren't fussy about quality since there was only one finished face. The very best of the flat slabs we saved for the top surface and for the steps. The broken pieces and unusable rubble were dumped behind the wall for fill.

Preparation

The first step is to lay out the base with string or mason's line. We had already set the levels for the top and the steps. The purpose now is to mark out a base area big enough to reach deep water, and small enough to need a minimum of concrete on the underwater pour. The logical line for the face of Tommy's dock still left some dangerous-looking rocks sticking up in the boat area, and left a steep slant under part of the base.

The bottom had to be re-arranged a bit. The big rocks beyond the string (in the boat area) were hauled up onto the shore for eventual use in the masonry. Cold work, but we consoled ourselves with the thought that rocks are lighter and easier to move underwater than they are on dry land.

Within the base perimeter, we had to level the bottom a little. In the first place, a slanty bottom might

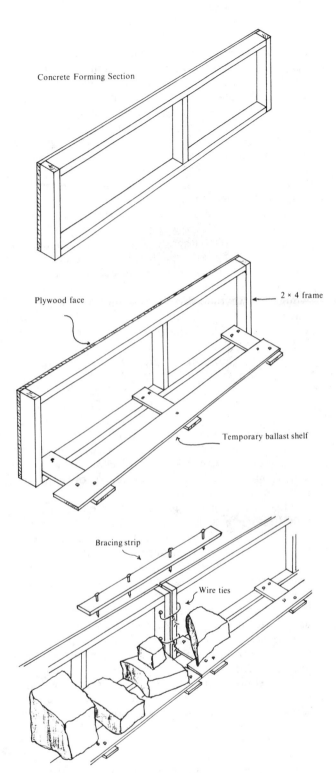

Figure 11-3   Getting the forms in place is one of the biggest jobs. First construct the forms themselves (top), add a ballast shelf (center), and join each section together with a bracing strip and wire ties (bottom).

121

encourage the big concrete slab to slide away into deeper water. Secondly, we needed a more level bottom to set the forms and weight them down. We weren't interested in leveling the solid stuff; it was the loose rubble that made the base unstable. An hour or two of rolling and dragging the worst of the rubble up onto the shore was sufficient. The result wasn't prairie flat — in fact, it was barely walkable — but it was good enough for concrete and forms.

The anchor rods came next. Every underwater crack in the ledge got a length of steel — hammered in as far as it would go. We varied the angles: some sticking out horizontally under water, others skewed up, down, or off to either side. Then more steel driven straight into the bottom, between the boulders. In most cases, we got at least 2 feet of steel into the rock, leaving 3 or 4 feet in the water where the concrete would go. Anything left sticking up was bent down under the waterline. If the ice ever moves this dock, it will have to break all that steel, or take the ledge and half of Tommy's lawn with it.

The biggest part of the preparation is setting the forms, and the first step is to put "feet" on the forms. Nail short pieces of scrap along one brace, and then one long board across the short ones. (Figure 11-3) This foot, or shelf, must be sturdy enough to hold a load of rocks, but don't be too fussy with the carpentry. This is a job you can measure with the eye, cut with a chainsaw, and bang together with common nails. In fact, a rough assembly at this stage makes dissembling forms easier at the other end of the job.

Float each form section into its approximate position. Leaving the string in place from the initial layout makes it easier to line up forms at this stage. Turn each form until the smooth side faces in towards the shore, and the shelf side sticks out towards deeper water. Now start piling rocks on the shelf. The form will sink lower and lower until you can maneuver it easily undermater. Move it into its exact position, aligned with the string, then pile more rocks on the shelf until it feels secure.

If the bottom is rocky, the form may still wobble unevenly. Use flat stones as shims, wedging and filling along the lower part of the form. The form should lean in towards the shore a little at the top, it should sit solidly on the bottom, and any gaps at the bottom should be plugged with stones. Any hole big enough for your hand is too big.

Now move the next form section into place and sink it with stones. Secure it the same way we did the first one. The two forms should butt, end-to-end. Rented forms may have holes through the bracing timbers. The holes are there to fasten the two forms together. Loop wire through the holes — on the outer side of the

forms so the wire won't end up in the concrete — and twist it tight. Greased bolts do an even better job, but are more awkward to line up underwater.

The two forms can now be braced into a rigid unit by nailing another board (3 feet to 6 feet long) across the tops. Overlap the joint and use at least three nails on either side. (Figure 11-3)

With a complete string of forms across the face of the pour, we floated the raft behind it and anchored it to the shore for extra support. (Photo 11-3) We also added a couple of vertical posts — driven into the bottom of the lake — some diagonal braces nailed to the raft, and wire ties between the forms and the reinforcing steel. At the time, it seemed like overkill; but the thought of what 20 tons of runaway concrete could do to valuable frontage made us cautious. When you think you have enough support behind the forms, double it.

Photo 11-3 *Anchoring the raft up against the forms will provide some additional support.*

The final task was encasing the Thomson's water supply line inside the 4 inch pipe and running it out through the form. Since we used perforated pipe for the casing, we covered the top with plastic to keep the concrete out.

The Base

Warn the concrete people that you're pouring into the water and will need a mix with very little "slump". In other words, not too much water! And ask that the mix be "air entrained." You will also want to prepare access, so the truck can back right up to the lake and dump directly into the form.

Pouring concrete into the water is not so foolhardy as it sounds. The concrete sinks to the bottom, displacing the water. Some dilution occurs at the interface between concrete and water. So the trick is to pour quickly, until the concrete rises above the waterline. When the forms are filled to the level of the lake, all the top water should be displaced. (Photo 11-4)

Any gaps at the waterline (open spaces at the ends, gaps between form sections, low spots where waves might come over the top) will add to washout problems. You will see underwater clouds of cement, drifting away with the current. Don't panic. Think of it as lime to neutralize the acid rain.

Washout spots can be minimized by plugging gaps in the forms, and by pouring quickly to displace the water before it dilutes too much. If the big gaps are plugged, the washout damage will be superficial.

In fact, behind the superficial rough spots, an underwater pour may be of higher quality than most home-made concrete jobs. The reason is simply that concrete must be kept wet for up to a week as it cures. Amateurs are more likely to ruin concrete by letting it dry out too fast. Think of the lake as a curing aid.

As the concrete fills the form, prod it gently with a spade to work it into all the little nooks and crannies below. Be especially careful to work it down beside the form. This will leave a smoother face on the concrete.

Don't skimp on quantity. Pour in enough to pass the waterline by an inch or two; this will help keep the

Photo 11-4   *After the forms have been filled with concrete to the level of the lake, all of the water should be displaced.*

123

first course of mortar out of the waves. Then add more concrete to slope the top surface slightly up and away from the water; this will help to drain the area behind the dock.

Prod the fresh concrete with a spade to work it into shape. Don't try to trowel or rake it smooth. As long as it drains properly, a rough surface is good enough. And it offers more "grip" for the first mortar course.

Finally, cover the surface with a sheet of plastic. The purpose, believe it or not, is to keep it from drying out. Then leave it alone for three days. Warm up, gather stones, go back to the office, but leave the concrete alone. By the time you get the mess cleaned up, the concrete may be hard enough to walk on. Don't be tempted to strip off the forms and get working. Wait.

Three days after the pour, we were back in the water, cutting the wires that tied the sections together, pulling up the stone ballast from the feet of the forms, and prying out all the braces. Naturally, it was another bitter morning, and all the more tempting to simply tip off the ballast rocks and let them sink. But we needed the stones for the wall, and it was some consolation to know that this was our last day of working in the water.

With the wires cut, nails pulled and ballast off, some of the forms simply floated free. Others came loose with the rap of a hammer. We stripped off the feet and carted the forms away.

The Wall

The face of the dock is built like a wall, but not like the wall of a house.

Remember that destructive forces are different in a dock. The wall of a house bears weight from above, but if the roof is well-balanced, the sideways forces are slight: the wind, slamming doors, that sort of thing. So the wall of a house can be quite slim in comparison to its height.

The dock, on the other hand, suffers most from the lateral forces; water pressure, the pounding of waves, ice expanding from the inside, ice shearing against the outside, boaters who can't find the brake. So the wall of the dock must be built to resist the forces that would tip it over, or push it out of place. It must be squat and solid, with a wide base and a smooth, sloping face.

The usual rules of good masonry apply (see Chapter 13), but for this job we'll have to add a few more:

1. Avoid the temptation to gain height quickly by setting a stone "on edge", veneer style. Every stone should be laid "flat", reaching as far back into the wall as it can.
2. Let the face slope back just a little towards the land. The mass of the wall itself doesn't lean, because the backside, too, tapers in as it rises.
3. Finish the face as smoothly as possible. That means turning the stone to expose the straightest edge; and it means flush joints rather than raked. If ice slides up the face and catches a projecting stone, the ice will always win. (Figure 11-4)

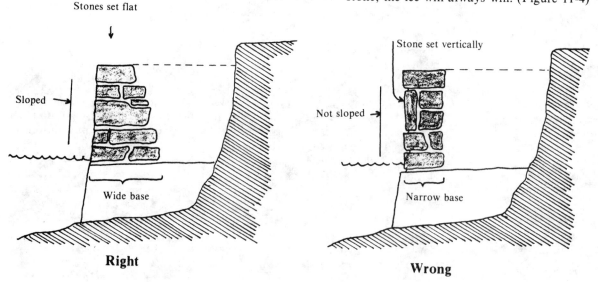

Figure 11-4   The proper method of constructing the wall is to have a wide base, the wall angled toward the shore, and the stones laid flat as the "right" drawing (left).

The stone face of Tommy's dock was to be 36 inches high. We wanted the first course — the base — to be that wide, so we cleared a 3 feet swathe along the front of the concrete and brushed it clean of all debris.

Since all the rocks would be laid "flat", we gathered a pile of slabs that were more or less the same thickness. This would ensure that when the first course was laid, the tops of the stones would be fairly even, making the second course easier to fit.

The mixer cranked out mortar as fast as Tommy could shovel in the ingredients:

9 parts sand
2 parts masonry cement
1 part portland cement

and then enough water to make it come out the consistency of mashed potatoes

Doug wheeled and shoveled the mortar over the ledge and dumped it on the wide base we had cleared. He slapped it down hard with the shovel to fill the rough surface of concrete, then he worked it out in a layer 1 to 2 inches thick.

Start the first stones at the face. Pick the straightest edge on the stone and slither it around until it lines up with the edge of the concrete. Then tap it down into the soft mortar until it feels solid. Fit the next ones beside the first, lining them up at the face, then fill in at the back. Like a jigsaw puzzle, each piece lies flat and fits (sort of) the contours of its neighbors. And, like a jigsaw puzzle, it's easiest to begin with the one straight edge.

The backside of this first course ended, more or less, at the back of the swathe of mortar Doug was spreading. That backside, destined to be buried in rubble, was the least of our worries. We were much more concerned with keeping the front face smooth and the tops even.

As the mosaic of the first course was taking shape, we stopped every 6 feet or so to place a length of plastic pipe between the stones. One end of a pipe fits flush with the stone face. The other end reaches back to the core — beyond the mortar bed and past the stones. This particular pipe had been coiled, and it still had a yen to spring up in curls. We, however, wanted it flat on the concrete base, to drain the core of the dock completely, without kinks and hidden pools. A few rocks, jammed over the top, will hold a kinky pipe in place.

At the back of the drains, inside the dock, we piled a small mound of loose stones over the end of each pipe, then laid plastic over the stones. The idea is to keep sand and silt from plugging the pipe after the core is backfilled. Water can seep around the plastic and through the rocks. The dirt is stopped by the plastic. (Figure 11-5)

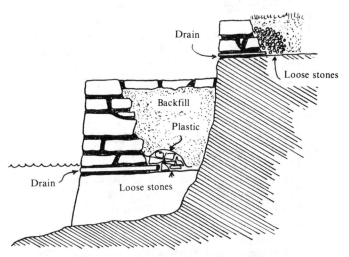

Figure 11-5   Good drainage for the dock is necessary. You should use drain pipe, loose stones and plastic to aid in this.

The second and subsequent courses go on just like the first: a layer of mortar, a mosaic of stones, front edges straight and aligned with the string, wiggle and tap each rock into a solid seat, fill the spaces between with small stones and mortar.

Each course gets a little narrower as the wall tapers towards the top. The sloping face is kept straight with the string. The backside wanders raggedly in and out as haste and imperfect shapes will allow. Rather like growing older, I suppose.

The last hour of each afternoon we saved for backfill and pointing. Backfill, to some extent, takes care of itself. The dross — rejected stones, dropped mortar, dirt kicked off the ledge as we climbed up and down — all ended up in the space between the ledge and the wall. We were careful to keep most of the organic trash out of the core, knowing it would shrink as it decomposed; but just about everything else got dumped in the hole. At the end of the day, Doug shoveled the dirty edges of the sandpile into the hole, until the hole was *almost* as high as the rising wall. We left the wall higher than the fill mostly to keep dirt out of the still-wet mortar. Walking and working on the fill each day helped to compact it in layers, a much more thorough compaction than could be achieved by filling in the entire hole at the end of the job.

While Doug shoveled, I climbed down on the raft with the last scoop of mortar and a small trowel to point all the joints done that day. Scrape off the excess that has squished out from between the stones, fill in the holes, and smooth the mortar until it's flush with the face of the stones. When it's hot and sunny, you might have to point the face every few hours. But in cool, damp, or cloudy weather, leave it until the end of the day. It's better to force a smooth surface on stiff mortar than to try to point too soon. When the mortar's too fresh, it won't stand up in the joint. No matter how much you work to smooth it, when you take the trowel away the mortar will slump away at the top and leave a tiny crack.

As the wall rose higher, we had to remember where the extras went: the bolts to hold the boat fenders, and the bottom boarding step.

The wooden fenders, or bumpers, are simply 4 × 4 timbers bolted to the face of the dock to keep the boats from scraping on the stones. Each timber is mounted vertically with two bolts, one high and one low. We planned to bury anchor bolts in the masonry, leaving 3 inches sticking out of the face. That allowed us to countersink the nuts below the surface of the timbers.

Somewhere around the second course of masonry, set four bolts in the joints between the stones. If the mortar isn't stiff enough to hold the bolt in position, jam in some little stone chips as wedges. Measure to be certain that the bolt doesn't stick out more than the thickness of the timber. And measure to keep the spacing of the four bottom bolts uniform. Don't worry about placing them all at exactly the same level. The timbers, after all, won't be drilled until the dock is done. But do keep the horizontal spacing uniform. Finally, if you have a grease gun handy, give the exposed threads of each bolt a slather of grease — it will make it so much easier to clean off the mortar drops before the nut goes on.

If all this sounds like a lot of trouble, you can leave off the timers entirely. After all, most boats fancy enough to worry about scratches carry their own bumpers on board.

Steps

We had already calculated the height of the bottom step when we measured the water levels. When the wall was still a foot or so shy of that level, we stopped to look for a big, flat stone to serve as the step. Better to cover a step with a single stone and avoid the need for toe-stubber joints on the surface. But there's another reason to scout out the step stone ahead of time: the thickness of the step stone determines the height and shape of the bed to be built under it.

Once we had found the almost-perfect step stone, we trimmed it here and there at the face, then measured the thickness. The stone was 6 inches thick. We allowed an extra inch for surface bumps and mortar, and called it 7 inches. The top surface of the step, remember, was designed to be 20 inches above the low

water mark (the boat deck was 12 inches higher than the water and 8 inches makes a comfortable step; 8 plus 12 = 20).

So the bed for the first step stone had to level off at 13 inches above the low water line (20 inches for the final height less 7 inches allowed for the step stone; 20 – 7 = 13). We set the mason's line at 13 inches and built the wall to exactly that height.

Those unfamiliar with the perversity of stone might ask: why not set the string at 20 inches and work to that level? The answer has to do with butter side down and never having exact change for the bus. When the height of the last course is critical, the stones are always the wrong thickness. If you leave yourself 4 inches for the last course, all the 3-4 inch stones will instantly disappear and you will be inundated with perfect 5 inch slabs. Believe me, measure the top stone *before* you prepare the bed for it.

With the bed at the right level, it was a simple matter to spread some mortar and set the pre-ordained step in its rightful place. Use the level to ensure that the top surface of the step shed water into the lake. If necessary, shim up the backside with thin chips to get the slope the right way.

Once the first step is in place, the rest of the wall goes up with whatever stones you can grab. At the second step, and at the top deck, we will again have to take care to get the height right: pick the top stones, measure their thickness, and compensate for that thickness in the underlying bed. But — for now — it's a simple matter of filling in around the step.

Whenever you're working up to a flat surface, like a step or a shelf, the easiest course is to build up the surface piece first, then fill in around it later. Doing the job in this order simplifies the hunt for the proper stone. Look at figures 11-6. First we got the surface right, then filled in around it, then built up to the next level. Fitted in any other order, the step stone would have to be square as well as flat, and we would have left continuous vertical joints.

The other steps — the ones that lead from the deck to the lawn — were built in the same order. First the steps, and then the retaining wall which incorporates them. Retaining walls are described in some depth in Chapter 7. But the steps had an interesting twist which is of interest here.

You'll recall that we marked the elevation of the deck on the ledge itself. From that line to lawn level was about 24 inches, or three 8 inch steps. As we dug away the sod and cleaned the brink of the ledge, one section appeared to be suitable for a bedrock step. There, the ledge had a clean, vertical face, and a straight, horizontal grain. It was a little too high for the first step, but the grain suggested that we might be able to cut it down to a flat surface.

Figure 11-6   To construct the steps you first select the step stone, then measure its thickness. Then set the mason's line at the bed height depending on the thickness of the step stone.

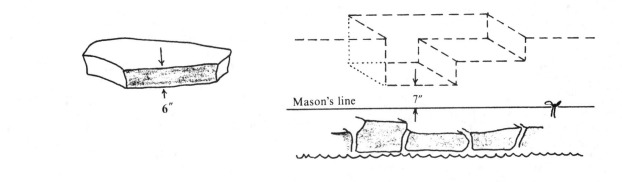

Mason's line   7"

6"

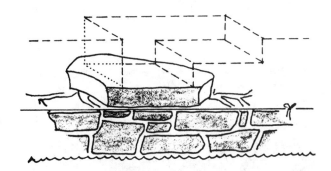

Prepare the bed, then set in the step stone.

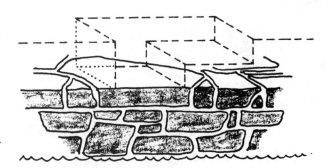

Fill in around the stepstone.

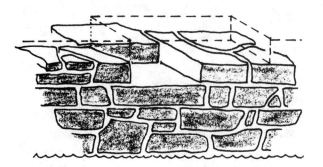

Start laying the stones on the next course.

We marked the top of the step, 8 inches higher than the deck line, and began to score the mark with a wide chisel. We worked back and forth along the line, cutting a long, horizontal groove in the sandstone face of the ledge. Tommy had to set up the grinder, to keep the chisels sharp, but eventually the groove was deep enough to take the steel wedges.

We hammered in two wedges, hitting one then the other in rotation. Within minutes, the groove opened to become a ½ inch crack, and the sound of the hammer changed from a high metallic ring to a lower note — a hollow "thunk." We tried the long steel bar, but it wasn't enough to pry the slab loose. The wedges were raising the slab far back under the lawn. Too far.

We marked a line where the back of the step should be, or the front of the next step, and hammered up and down the line with the sledge. The top slab snapped along the line of hammer blows. We had a natural step. (Figure 11-7)

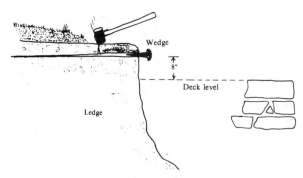

Figure 11-7 We were able to form a natural step-stone by wedging out a large slab section from the natural bedrock at the ledge.

We mortared in flat slabs for the next two steps, remembering to include a length of ¾ inch hose to drain ground water away from behind the steps. Only after the steps were in did we scrape off the rest of the ledge and build the low retaining wall. Needless to say, we put lots of loose stone behind the wall, and drains to let the water through.

Deck

Meanwhile, back on the dock itself, the wall rose closer to the final elevation. At this joint we paused to gather a truckload of heavy slabs ... just as we had paused earlier to find and set the step. Here, our concern was to top the wall with a solid, uniform course of heavy stones. An inch too high or an inch too low wouldn't be good enough. First, variations in height are surefire toestubbers. Secondly, we could not expect to level the top with thin "filler" stones. The

stones that top the wall, called "capstones", must be heavy enough to hold themselves in place. The waves, the banging of boats, shovels, brooms and feet would soon have a lightweight rock knocked loose. Finally, the top of the wall is too easily compared to the perfectly level waterline. With every passing sailor eyeballing the edge to see if it's straight, these capstones are worth the extra effort to get the best.

Doug, undaunted by the trip from the concrete plant, put his trusty truck in its walking-on-water gear and launched it from the nearest beach. There's a shoal, you see, where the rock is ideal for capstones. It wasn't even that far underwater — one inch higher and the water would have snuffed the engine, one inch lower and it wouldn't have poured over the tops of my boots. Just right, according to my amphibious friend.

Since all our capstones came from the same shelf, or stratum, of sedimentary rock, they were a uniform thickness: about 5 inches. So we set the mason's line 6 inches lower than the final elevation (5 inches of capstone and 1 inch of bumps and mortar) and built up the wall to the line.

Photo 11-5 The top and bottom anchor bolts should be aligned exactly for easy mounting of the bumper timbers.

This last stretch of all beneath the capstones holds the top bolts for the wooden bumpers. Bury the anchor bolts just as you buried the lower bolts. This time, however, take care to align the top bolts exactly with the bottom ones. (Photo 11-5) If each pair of bolts isn't plumb, the timbers can't be mounted straight, and some smart-aleck sailor is bound to point this out. If this bothers you, leave off the bumpers and let him provide his own. Serves him right.

With the wall at the right height, it's a simple matter to slather a bed of mortar on top and set the final capstone course. As you set each stone, check it with the level to ensure that the top surface is perfectly flat, or sloping slightly towards the lake.

The capstone course is part of the deck, and the deck has to be able to shed water — not trap it.

When the capstone course is complete, and before the mortar sets, paddle off into the lake for the sailor's eye view. In particular, look at the horizontal line of the top course and compare it to the water line. They should be perfectly parallel. If there is a wiggle or bump, this is your last chance to fix it.

Do resist any temptation to mount cleats or tie rings in the capstone course. The forces exerted on these are considerable, and the capstones are too vulnerable. Better to mount the tie rings farther back from the edge — somewhere in the deck itself.

At this point, we topped up the backfill and turned our attention to the retaining wall at the back of the dock. (Photo 11-6) There are two good reasons for this: first, we wanted to compact the top layer of fill by tramping around on it for a few days; and, secondly, we didn't want to spoil the flagstone deck with spatters of mortar from the wall above.

By the time we got back to the deck, there was a tinkling skim of ice on the lake. We set the flagstones on the sandy fill, leveling them roughly as we fitted. In truth, there wasn't much left to be done. The area around the steps, and the capstone course, finished much of the top surface. We started with the biggest flagstones in the widest gap, and worked our way back to the final fitting around the derrick and the lamp post.

All that remained was the final leveling, then working mortar into the surface joints. Those two jobs we left until spring. It wasn't the weather that chilled the workers' zeal. It was the odds on the backfill settling some more. By spring, the lake would be up to the capstone course, and the fill would be saturated with water. The water would wash the sand into any tiny nooks that our trampling might have missed. Had we not had the spring flood to do it for us, we would have soaked the backfill thoroughly with the hose to help compact it.

So Doug took Wondertruck hunting, and Marion took Tommy to Florida. The rest of the deck waited for spring.

Photo 11-6   Once the wall has been finished you are ready to finish backfilling in the soil, and finish off the surface deck.

Chapter 12

Super-Hibachi

I'm sure the poor man meant well. He came roaring up the drive at great speed, shouting that our outhouse was on fire. We thanked him for his concern (just a little coldly), and advised him that IT was not an outhouse and IT was not on fire. There was nothing inside but the Christmas ham, and one should not be alarmed at the sight of smoke curling out from the eaves of an ad hoc smokehouse.

We were, nonetheless, taken aback by the breathless stranger. We had to admit that, yes, it did look like a chipboard privy. And perhaps it was time to spruce IT up, whatever IT was.

IT had evolved from a hole in the ground, a pit for summer cookouts, and a place for every other kind of fire that a country home needs from time to time. Now it serves as smokehouse, maple syrup maker, bake oven, summer kitchen and barbeque. It is not, never has been, and never will be . . . an outhouse. We call it Chateau de fumée, or Super-Hibachi.

Practical people scoff at building such an elegant structure for those lowly chores. And maybe it is a little too ambitious in a book of modest, backyard jobs. But there's a reason for having it here, and a reason for putting it last. It's the exuberance factor.

The fact is that amateur masons are serious folks — solid, four-square, sensible people, respectors of craft as something other than cheese. These are people who build as if the world will outlive them, people who never trifle with shoddy short-cuts, kits, and commercial nonsense like (. . . arrgh . . .) pre-cast blocks. Expecting a stone mason to build something fancy just for the fun of it is like waiting for Martin Luther to break into giggles.

But there is a joy in learning any craft, and masonry is no different. We start with modest little things that can't go wrong but do. Then try again when no one's looking, and surprise ourselves with how good it feels. Eventually, joggers enter marathons, dancers dip, and backyard builders plan the ultimate project — something guaranteed to wow 'em at the fence. "Gawd, Herb, that's a beauty. They don't makeem like that anymore . . . whatever it is. Lost art, eh!"

And that's how we got to Super-Hibachi from a lowly little smokehouse that looked, for all the world, like a smouldering privy. You might not want a smoke in the backyard, but I know that after you've done a few of the more modest projects you'll want to try a masterpiece of some kind — anything to stretch the new-found skill to its joyful limit. Go ahead and laugh, Martin Luther, but that's the way it goes. And if you don't need a Chateau de fumée, I'll tell you later how it can be converted to sauna, garden shed, or the world's prettiest incinerator. Just to get started, though, let's assume that some readers might actually want to do a little home smoking.

Design

The trick is to get the smoky flavour into the meat without actually cooking it. If the smoky atmosphere gets too hot, the fat begins to melt, the lean dries out, and the half-cooked meat is prone to spoil. 75-80° F is about right.

The easiest way to cool the smoke is to separate the firebox from the smoke chamber, and run the connecting smokepipe through a cooling medium. A long, metal chimney, exposed to a cold winter's day, will do the job, but the smokehouse temperature will rise and fall with the sun and the wind. The more reliable way is to use a heat sink: run the pipe through the earth, or through a mass of masonry. The masonry soaks up the heat and moderates the inside temperature.

There are lots of ways to build a smokehouse. This was to be our third. This time, we wanted one big enough to do a year's smoking at a time, and designed so that we could leave the fire unattended for a day without it going out, and without exceeding the critical temperature. Most of all, we wanted it to look nothing at all like an outhouse.

The heart of the plan is the firebox. It needs a proper door, with an operable draft to control a slow burn. The box should be lined with firebrick, and surrounded with a generous mass of stone, to soak up the excess heat.

The smoke chamber needs a pressure vent to let the smoke in easily. Without it, smoke would leak from the firebox, but back pressure in the smoke chamber would keep the smoke out from the one place you want it. The idea is to exhaust air from the bottom of the smoke chamber, leaving the smoke and the ham near the ceiling. Even after the fire goes out, smoke will hang in the upper strata for hours. (Figure 12-1)

Even with a pressure vent, when the fire starts the smoke won't know which way to go. It's as easy to put her back out through the firebox door as to find its way through the smoke chamber. The answer is an ordinary chimney to get the fire started and the air currents flowing in the right direction. Once the fire is established, and you're ready to start smoking, close the main chimney and open the flue to the smoke chamber. The other advantage of this approach is the option of starting with a roaring fire to build up a day-long bed of coals. The heat gets diverted out the main chimney, so the smoke chamber doesn't get too hot. And you can start the fire with any old scrap, saving the aromatic hardwoods for the flavouring phase of the burn.

With memories of winter rain trickling down the back of my collar, I planned a roof that would be bigger than just the smokehouse. It would have to stick out at least far enough to cover the firebox and the fireman. And so, it might just as well be big enough to also cover the woodpile and barbeque supplies.

And then (as long as we were building a central chimney and an extended roof anyway) why not

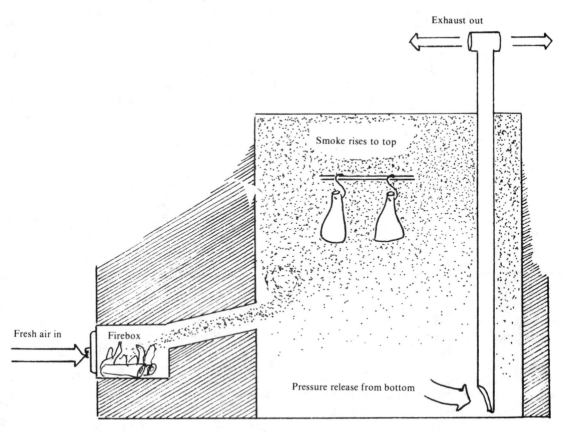

Figure 12-1   The smoke chamber will need a pressure vent to let the smoke in easily, and a low level exhaust.

Photo 12-1   We also decided to incorporate a maple syrup maker into the smokehouse design.

include the maple syrup maker? (Photo 12-1) And April's sap boiler could be August's barbeque pit. And the waste heat from the smoke might warm an oven. And so it goes. Do you see what Byzantine complications can arise when you build for the fun of it?

With all the extra functions added to a simple smoker, the innards — the flues, diverters and fireboxes — got a bit complicated. My usual "blueprint", a rough sketch on squared paper, began to assume Frankenstein twists. The sketch also revealed several violations of the first principle: smoke rises! Smoke, according to those first rough drawings, would have to stoop and twist like a politician if it had any hopes of emerging from the proper hole. The critical task, at the planning stage, was to check the elevations of all flues and openings to ensure that their angles were always up and their outlets were higher than their inlets. The amended, final plan is shown here. (Figure 12-2)

That was the hard part. The easy part was the whimsey. The innards might resemble a nuclear reactor, but the exterior would be more benign. It wasn't entirely frivolous. The flower box atop the end wall would replace 12 square feet of hard-to-find capstones. (Photo 12-2) The log mantle over the oven was easier to cut than a big, stone mantle. And the cupola would help the clouds of maple steam and barbeque smoke escape through the roof. All very functional. But, to be honest, they first went into the plan because they looked good. That's the way it goes.

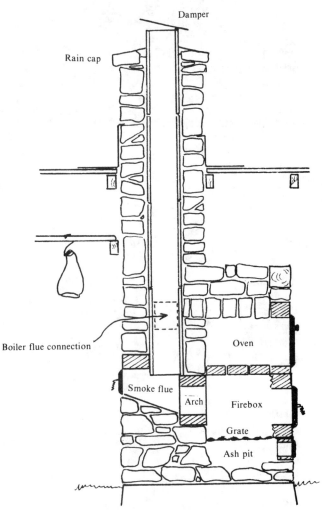

Figure 12-2   The amended, final plan of the smokehouse is shown above.

133

Photo 12-2   The flower box above the end wall saved finding and setting 12 feet of heavy capstones.

Materials

Stone, for walls and firebox exterior
Concrete, for footings and pillar caps
Sand and cement
Ash grates — 2
Firebricks and high-temp mortar
Precast chimney section, for interior arch
Chimney brick, for openings
Flue tiles
Angle steel, for grill/boiler
Cast iron fire doors — 2
Door hinges or scrap to make them
Oven door
Small clean-out doors, for ash pits — 2
Large clean-out door, for smoke chamber flue
Flashing metal
Posts and timber, for roof
Sheathing and shingles, for roof
Framing lumber and door, for smoke chamber
Plastic pipe and rubber flashing, for pressure vent

The only criterion for the stone is to remember that it will be subject to extremes of heat. We've protected the fireboxes and flues with heat resistant liners, but the stone claddings will still have to endure temperature fluctuations that would crack poor quality stones. Avoid anything that already shows fracture lines or other weaknesses.

The ash grates can be anything that comes to hand. We used part of a cast-off coal grate from a fireplace, and somebody's home-built steel barbeque grill. It will function as long as the spaces are large enough to shake the fine ash through, and small enough to keep the chunkier, live coals up in the firebox. If you have a choice, cast iron is better than steel (which can warp at high temperatures).

Firebrick is essential for lining the fireboxes and flue openings. We used thirty-six 2 inch bricks, and twenty-four 1 inch bricks. We also used three large bricks ($18 \times 6 \times 2$) for the bottom of the oven. The large bricks are harder to find — these were salvage from an industrial furnace. If you can't obtain them, some alternatives are suggested at the end of the chapter.

High temperature mortar is made specifically for laying firebricks. It comes pre-mixed, in cans, and a one gallon can should be plenty.

I used a pre-cast chimney section for the interior arch, between firebox and flue. These are available from building supply stores in a variety of grades and sizes. Remember, though, that a chimney block was not designed for use as an arch. In effect, we're going to stand a section on its side and use it to support the chimney. Do not try to substitute something as flimsy and fragile as clay tile. Use the heavy grades, or make up an interior arch out of firebrick.

Firebricks may not take the pounding of an iron door, so we framed the firebox openings with tougher, chimney brick. Not all brick can stand the heat, so ask for chimney brick. We also used some salvaged chimney brick for the domed top of the oven, but temperatures are lower that far from the fire, and stone would work as well as brick.

The main chimney is lined with clay flue tile, four 8×12 sections. The short, diagonal flue, from the boiler firebox to the main chimney, was cut from a single section of 8×8 tile. Each tile section is 24 inches long.

Our evaporator pan slides over the fire on steel rails, cut from the frame of an old bed. The rails also hold the barbeque grill. Any length of angle iron will make a rail, but if you don't need a sliding pan, a brick-lined top would hold the barbeque grill.

Cast iron doors for the two fireboxes are functional necessities. Having draft controls built into the doors makes them even more useful. You can buy doors from furnace or woodstove suppliers (doors are sometimes sold as part of barrel stove conversion kits). However, it is considerably cheaper to find used doors. We found one in a second hand store, and the other in a scrap heap behind a local foundry.

The only drawback to junk doors is the inevitable missing hinges, latches and handles. A small bedspring makes an ideal handle. Hinges and latches can be improvised from scrap iron. (Figure 12-3)

An oven door is easier to find than to buy. An ordinary cast iron door, like the ones we used for the

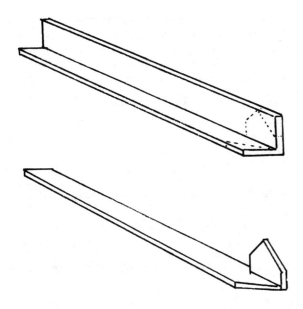

Figure 12-3   A simple latch made from angle iron.

fireboxes, will do; but temperatures are not extreme at the oven level, and other possibilities arise. We used the door from a discarded stove. A home-made door would do as well.

The small doors for the ash pits are cheap and readily available at building supply stores. Also called "clean-out" doors, they are commonly used at the base of a chimney. The doors come complete with a metal frame that can be mortared right into the masonry. Or, they can be set on hinges, as ours were. A larger "clean-out" — one with a hinged door — makes a handy control for the smoke flue, inside the smoke chamber. Closed, it diverts smoke up the main chimney. Open, it allows the smoke into the chamber.

Flashing metal comes in rolls, in aluminum or galvanized, and in various widths. Aluminum is easier to shape than galvanized. A 20 inch width is sufficient, and you'll need about 24 feet of it.

Lumber to frame the chamber and roof is a matter of what's available. We used 6 × 6 cedar posts and cedar framaing only because a local mill had a glut at the time (and they threw in the firewood slabs for nothing). The main door to the smoke chamber was another discard. We framed the doors, then laid the stones around it, solving the problem of a door to fit the space by making the space to fit the door.

3 or 4 inch plastic pipe, sold in the plumbing department, makes a fine pressure vent. You'll need enough to reach from the ground to 2 feet higher than the roof. In this design, that's an 8 foot pipe at the low end of the roof. An ordinary "T" junction makes a rain hood, and a rubber vent flashing seals it at the roof line.

Preparation

A building this size needs a solid base. You can put down footings and perimeter drainage, as you would for a conventional building. Or, use a concrete slab on grade. In any case, local climate and soil conditions are the first considerations.

Here, where bedrock lurks a spade's depth under the sod, we simply excavated footings, cleaned off the bedrock, and poured concrete into the shallow earthen "forms".

We set anchor bolts around the inner perimeter of the smoke chamber, but only because we intended to frame the smokehouse and test it before committing the structure to stone. If you intend to go straight to a solid stone wall, without the frame interior, skip the anchor bolts and the framing. You will have to put top plates on the stone walls to carry the rafters over the chamber; and you will have to frame the door separately and "key" it into the masonry. Otherwise, the procedure is much the same.

Start with a course or two of stone around the firebox base. At ground elevation, there are no flues or cavities to worry about, so use the big, irregular stones here and fill the core with rubble. You can start the ash pits at any level, but we chose to raise the base masonry about 6 inches first, mostly to keep the clean out doors higher than the spring slush.

From this point on, elevations are critical. Getting all the cavities and components to come out at the right height requires starting them at the right height. Here, it seemed that every few inches of elevation required a beginning for some other part of the system. The safest tactic is to have a detailed plan of the interior, sectional drawings with elevations, and post it at the site. At every course, check to see what comes next and at what level it starts.

The two ash doors come first. We framed the openings with firebricks. The bricks are the right size (4½ × 9) to make a flush surround for the door. At this level, away from the direct heat, fit the bricks with ordinary mortar.

The surrounding masonry has to hold the door hinges. For these light doors, we found that pre-drilled steel mending plates (available at any hardware store) hold the hinge pins nicely. (Photo 12-3) Round off the corners with file or grinder if they interfere with the door.

To make certain that the hinge holders are in the right alignment, mortar them in, hold the door in the closed position, then wiggle the pin holders around in the wet mortar until they hold the door firmly back against its firebrick frame. Set a stone over the top hinge to keep it from popping out of the mortar. When

Photo 12-3   The hinges for the firebox door were set right into the concrete on the sides.

everything is in position, carefully lift off the door, or prop it in position with a board. Leaving the weight of the door on the hinges at this stage will pull the hinge holders out of the still soft mortar.

The ash grate is just above the door in elevation. It covers the ash pit, but it doesn't have to cover the whole area of the firebox. Indeed, it must not be larger than the firebox. If the grate is wider than the firebox, then some part of the firebox masonry will rest on the grate. That is to be avoided for the simple reason that metal expands at a different rate than the masonry. Expansion and contraction from the heat would weaken the masonry at the point where the grate was imbedded in the mortar. By all means set the grate on a mortar bed if that will help to keep it in place, but do not rest the firebox walls on the grate. (Photo 12-4)

Beside the grate, where the firebox walls will be built, level the top course of masonry carefully. This is the base for the firebrick. And the firebrick, with its narrow joints and precise dimensions, demands a straight and level beginning. If there are undulations at the base, the firebrick wall won't work. If your stone is too rough to level the base, use brick, or small aggregate concrete to create a perfectly flat ledge.

Fireboxes

The high temperature mortar is not at all like the thick, forgiving mix that stone masons are accustomed to. It's buttered onto the thin joints like glue, and it dries quickly. The trick is to know what goes where ahead of time, and work quickly.

Take a dry run at the firebox first, stacking the bricks without the mortar. Staggered joints will demand at least some cuts, and these can all be made ahead of time. Firebrick cuts easily with the wide chisel; or, to be fancy, use the masonry blade in a power saw.

Now is the time to plan the interior arch — the passage between firebox and chimney flue. The photograph shows two ways to do this, one with firebrick and the other with a pre-fabricated chimney section. (Photo 12-5) Generally, the arch has to be strong enough to carry the chimney's weight, and heat resistant to keep from cracking. The arch can be placed anywhere around the firebox, but do try to place it higher than the fuel door or the stoker will get a faceful of smoke. Ideally, the fuel door is at one end and low; the arch is at the other end and high.

When you're satisfied with the dry-stacked version of the firebox, when the door hole and the smoke hole are in the right alignment, and when all the cuts are straight, mark the base with chalk (so you can put it back exactly the way it was) and dismantle the bricks. Soak them thoroughly.

Using a small trowel, or even a putty knife, spread the mortar along the base and set a corner brick. Tap it down into the mortar and level it carefully. This high temperature mortar is softer than the usual stuff, and will settle down to a joint of ¼ inch or less. Now butter the end of the next brick, the one that abuts the first, and tap it into place. Again, check the level and the alignment.

When the first course is complete, level and aligned, start the second. Don't worry about cleaning up the

Photo 12-4   Build up the firebox wall around the grate as shown in the photograph.

Photo 12-5   *The two fireboxes show two ways to design the interior arch: on the left the arch is built with firebrick and on the right with a pre-fabricated chimney section.*

joints. Just get the bricks straight. The second and subsequent courses will have to be checked for vertical alignment as well as horizontal. Move right along, though, or the mortar will lose its smooth, elastic texture.

When the topless box of firebrick is complete, strike off the excess mortar with a trowel, and cover the whole thing with plastic. The plastic keeps the dampness in for curing, and it keeps the showers out.

The next step is to add the arches. Level the top of the arch carefully, since this is part of the base for the chimney tiles. And, again, use high temperature mortar — the arch is often the hottest part of the entire system.

The smokehouse has its own arch — from the chimney base to the smoke chamber. This passage is cooler than the firebox arch, and so it can be built with chimney brick. Slope the bottom so it rises from under the chimney, up to the smoke chamber door. Add the sides, but don't cover the top of the passage until later, after the first chimney tile is in place. At this stage, when the passage has a bottom and two sides, mortar in the door that allows smoke into the chamber.

Now, prepare a level base for the first section of chimney tile. The firebox arch forms the front ledge. Opposite that arch is the open-topped passage to the smoke chamber. Build up either side with chimney bricks, bringing the tops exactly level with the top of the firebox arch. The result should be a three-sided ledge, level all around and shaped to hold the 8 × 12 chimney tile. (Photo 12-6).

We angled the flue from the other firebox (the boiler/grill) into the first chimney tile. This connection should be higher than the smoker ducts. To make an angled connection like this, start with the smaller 8 × 8 tile and cut it at a 45° angle. *Where* to cut depends on the separation of the two flues to be joined. A 4 inch separation, for example, requires you to start the cut 5 5/8 inches from the top of the smaller tile. (Figure 12-4)

Cut the top off the smaller tile and reverse it. That is, put the square end against the angle of the bottom piece, and put the cut end against the larger tile. Mark the larger tile and cut out this hole. Now the connecting piece can be cemented into place. The two pieces of smaller tile will not fit together neatly because the

Photo 12-6 *The base for the first chimney tile should be build up to the top of the arch, be three-sided with an opening to the smoke chamber, and be level so that the chimney is exactly vertical when set down.*

angled cut is longer than the square end of the tile. The overlap, however, is not significant. Slather some high-temp mortar over the cracks, smooth the insides of the joints by reaching in from the top, and leave it to set.

Cutting tile is a simple, but unpleasant, job. Mark the cuting line, then score it using a masonry blade in the power saw. It may take several passes, sawing a little deeper each time. Use patience and a gentle touch, or you risk breaking the blade as well as the tile. The unpleasant part is the dust. Wet the tile as you cut, and use the best dust mask you can get.

Now wrap the whole fragile-looking firebox in a cladding of brick and stone. The brick goes around the loading door, and over the arches, where the heat may be too intense for stone. Don't put the stone cladding too solidly against the firebrick — leave some room for expansion. A touch here and there won't hurt, but don't try to fill the gap completely with mortar.

Where brick and stone meet — around the doors, for instance — overlap the two materials course by course. If the stone is too big, or too irregular, to mesh easily with the brick courses, use metal "brick ties" to join the two materials. Place half of a tie between bricks, bedded in the mortar joint. When you abut the stone to the brick, bend the free end of the tie to fit between stones.

Before you start bricking around the door frames, measure the doors and decide on a proper height to set the hinges. The firebox doors are heavy, and the hinges

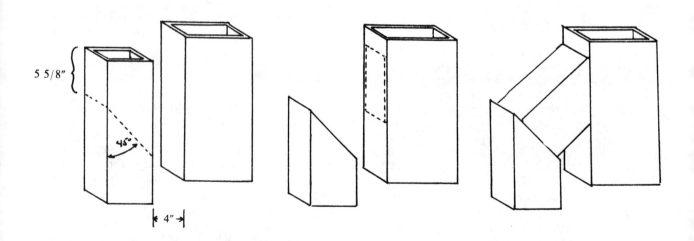

Figure 12-4 *To make the angled connection between the first firebox flue and the chimney, you must connect it at a 45 degree angle.*

must be solidly set. (Photo 12-7) I used pre-drilled steel scraps in a variety of shapes; generally, anything flat enough to fit between the bricks and sturdy enough to hold the door, will do. Remember to also brick in the catch on the other side of the door opening. (Photo 12-8)

Photo 12-7   The hinges of the oven doors are long enough to be set in between the brick layers.

Photo 12-8   Brick in the catch on the other side of the oven door as well.

There's a trick to getting a flush fit between the bricks and the door. Brick in the entire door frame, including hinges and catch. Then — while the mortar is still soft — carefully set the door in the hinges and press it closed. You will have to hold the weight of the door since the hinges can still pull out of the soft mortar. When the door looks straight, prop it in place. Then gently tap the bricks up against it, until the fit is tight all around.

Close the gap over the top of the door with a brick arch. I used 1 inch firebricks, cut in half and set on edge with their good ends out against the door. First, make sure there is a solid masonry abutment at each end of the arch. Then cut some 2 inch styrofoam to fit in the door opening. Cut the top of the foam where you want the bottom of the arch to be. If you want to be fancy, you can give the top of the foam support a slight, freehand curve. Now start at an abutment on one end and mortar a brick against it. Mortar the side of that brick and tap another one against it. Fill in the entire arch with bricks and mortar, then tap them out until they all touch the door, the same way we aligned the bricks around the sides of the door. Leave the foam support in place until the next day.

I capped the firebox with 18 inch firebricks, set behind and flush with the top of the arch over the door. Large firebricks may be hard to find, however. The best alternative is to cap the firebox with a firebrick dome.   Build a dome cap in the same way I capped the top of the oven (next section). If you are going to use a dome cap, build it simultaneously with the arch over the door, and weave the two together with overlapping joints.

The boiler/grill sports a topless firebox. When the boiler pan is in place, the pan acts as a cap, and the smoke is drawn under the pan and up the chimney as it would be in any wood stove. When it's an open-topped grill, however, smoke up the chimney is a matter of luck and the breezes. The smoke doesn't harm the stonework, but the heat can. That's why I lined the rim with 1 inch firebricks. They protect the masonry, provide a neat cap, and cover the gap between the firebox and the stone cladding.

Oven

The oven was a bit of whimsey that began as a wish to build a domed cap. The oven door was built-in just like the firebox doors: we bricked the surrounding frame, setting the hinges in the brickwork, then used the door itself to straighten the bricks for a flush fit. The side walls of the oven were stone.

A dome is an arch, and arches push out at the sides. They want to widen the gap beneath them so they can fall. This means that we couldn't put the dome cap on until the side walls were thoroughly set and solid. Then the fun begins.

Trace the desired shape of the arch on a piece of 2 inch styrofoam. There's no need to be fussy — any slight curve will do as long as it is symmetric. Cut out the shape with a kitchen knife, then make two more pieces exactly like the first one. Stand these up in the oven space and bend a piece of sheet metal over the curve. Cut the sheet metal so that it goes *almost* to the side walls. If it rests *on* the side walls it will be hard to remove afterwards.

The foam props and their sheet metal top make a supporting form for the arched dome. The dome may be brick or stone, but the uniformity of brick makes a tricky job easier. (Photo 12-9)

Photo 12-9   The arched dome over the oven is more easily formed with uniformly shaped bricks.

First, set up the bricks without mortar, just to check the fit. Start on one side with a course of bricks, all set on edge. Leave enough space between them for the mortar joint. Now add the second course, again setting the bricks on edge and leaving narrow gaps between them. Start this second course with a half brick beside the full brick that began the first course. This staggers the joints and strengthens the dome. Continue course by course until you approach the centre line.

Now start at the other side wall and lay out the other half of the dome. If you're lucky, there will be a gap left at the top, and the gap will be just a little bit more than one brick wide. If the gap is close to one brick wide, you may be able to adjust the width of the mortar joints to come out right. Otherwise, you will have to find some skinnier bricks, or cut them.

When the fit is right, dismantle the bricks and rebuild the dome. This time use mortar. As always, wet the bricks. Work quickly to complete the dome before the first of the mortar begins to set. When you reach the gap at the top, you may need to tap the bricks into a tighter fit, in order to squeeze that last row of bricks into the top, key course.

After all the bricks are in place, work more mortar into the joints to ensure there are no hidden gaps.

Now comes the fun part. Reach in through the door and *gingerly* wiggle the foam props loose. Cut them if you must. Pull out the props and peel the sheet metal off the ceiling of the dome. The dome may drop a tiny fraction, as its weight squeezes the mortar tighter, but it should not fall. Do be careful, though, not to push any bricks *up* as you remove the props. It is the fact that they are all trying to drop simultaneously that keeps them in place.

That's the fact, but there still seems to be some magic in an arch, some element of unnatural levitation. If you're still a white-knuckle mason, go have a drink before pulling out the props. But don't leave it too long. The mortar that has squeezed between the dome and the form will be unsightly if allowed to set hard. Pull out the props while there is still time to strike off the excess and smooth the joints in the ceiling.

Walls

Once the fireboxes, flues and caps are done, the rest is ordinary stone masonry. Lay up the walls in the usual way: stagger the vertical joints and keep a straight face.

If you're building around a pre-framed structure, like our chipboard smokehouse, finishing the top of the wall at the low eaves can be a little tricky.

If there is no roof sheathing in place, start the sheathing at the eaves and bring it up the rafters, past the line where the stone face should meet the roof boards; then build the masonry up under the sheathing by reaching down through the rafters from above.

If the roof is already covered, nail blocking between the rafters where they should meet the stone face. Then build the masonry up to the blocking by reaching around behind from the adjacent rafter space.

It this were a habitable structure, it would have been worth pulling off the chipboard and building the wall without backing. Moisture will eventually rot the chipboard where it meets the stone. When that happens to the smokehouse, however, we'll just pull away the rotten chipboard and get along with the stone wall behind it.

As the wall rises, remember to build-in the ends of

the beams that will carry the rafters over the woodshed portion of the structure. Since the other ends of these beams rest on posts and pillars, leave the wall at this point and go to work on the roof supports.

Roof

The four posts rest on stone pillars for three very logical reasons: first, to keep the bottoms of the wooden posts up out of the mud; second, because I happened to have four short cedar posts and no long ones; and finally, I just thought it looked better that way. Similarly, the low wall at the end is a wall and not two pillars because I had some very large stones left over from the last project; and the stones were too big for pillars.

At any rate, the pillars and the wall rest on concrete pads. And they are capped to shed the rain. Water lying on the top of an uncapped wall will deteriorate the masonry. You can pre-cast caps on the ground, or pour them in place.

Pre-casting takes a simple box form, and a not-so-simple lift when it's ready to go on the wall. Mortar the finished cap on the top of the pillar as if it were one big stone.

Forming is much more difficult when you pour the cap in place. (Photo 12-10) The drip edge has to extend past the stone face; so the form has to be propped up, and the gap plugged at the lip (see the forming for the gateposts in Chapter 3).

Photo 12-11   To anchor the wooden posts set spikes in the concrete which will hold the posts.

Photo 12-10   The wood form in place for the pouring of the concrete capstone.

No matter how the caps are poured, make some provision to anchor the wooden posts atop them. It would be possible to form a shallow mortise in the cap, a recess in which to fit part or all of the post end. This, however, would collect water and eventually rot the wood. We set spikes, pointy end up, in the wet concrete. (Photo 12-11) When the cap was in place, we put a square of roofing paper (as a moisture barrier) on the end of the 6 × 6, then drove the post down on to the spikes.

The beams cross the posts, and are mortared into the smokehouse wall. Here, the diagonal knee braces are neither whimsey nor decor. The freestanding posts would wobble without these braces. And the braces shorten the effective span of the beams. Without them, the 4 × 4 beams would sag over that span.

The rafters come next. I assembled simple "A" trusses on the ground, then lifted them up on the beams. Set the rafters before completing the masonry walls. This order allows you to bring the top of the masonry up through the rafters, working from above.

When the rafters are up, finish the wall, and then sheathe the roof. Leave the pressure vent and the shingling until you're ready to put the base flashing around the chimney.

Chimney

Before raising the chimney beyond the first tile, cover the area at the bottom of the chimney with paper. Any mortar that drops down the flue lands on the paper. Pull out the paper when the job is done and the mortar mess is gone. Just like a bird cage.

Cement the tiles together with a straight portland and sand mix (1:3). Cementing the tiles takes some special care. First, chimney tiles can soak up water like a long distance camel. They'll suck the mortar dry before you can get the tile plumbed. Soak the tiles in a bucket before you begin.

Lay a ridge of mortar around the top of the first tile, then set the next section onto it. Don't wiggle it down into the mortar until you've plumbed it. Check the vertical alignment with the level. Then *gently* tap the high side with the wooden end of the hammer. Recheck the vertical and tap some more until it's straight. If the mortar is too stiff, or too thin, to allow the tile to settle into alignment, pull off the tile and try again with fresh mortar.

When the tile is plumb, leave it for half an hour to let the joint set up. Then reach in from the top and clean off the excess mortar that squeezed out of the joint. A ridge of excess mortar outside the tile won't do any harm, but inside the tile it encourages creosote build-up and makes cleaning difficult.

A chimney rises a tile at a time. First the tile, then the stone to cover it, then another tile. The stone cladding should not be tight against the tile. Leave a narrow gap for expansion, and try not to drop too much mortar down the gap. Keep the tiles on the vertical by aligning each one with the one beneath it, and checking the sides of the tile with the level. Keep the stone cladding vertical by checking each course at the corners — drop a plumb bob from the corners to the chimney base. If the tiles are plumb, and the faces of the stonework are plumb, then the thickness of the masonry around the tile will be constant. And that is the extra check: if the masonry becomes thicker or thinner on any side, then either the tiles or the masonry must be wandering off the vertical. (Photo 12-12)

When the chimney passes through the rafters, remember to leave a gap of at least 2 inches between the masonry and the wood. Then cover the gap with metal flashing.

Chimney flashing consists of base flashing and counter flashing. Base flashing attaches to the roof but not to the chimney. Counter flashing overlaps the base flashing, and it attaches to the chimney but not to the roof. (Figure 12-5) The exact pattern of the pieces depends on the pitch of the roof.

Photo 12-12   Set the chimney tiles vertically true as you rise up, setting the stone cladding around it.

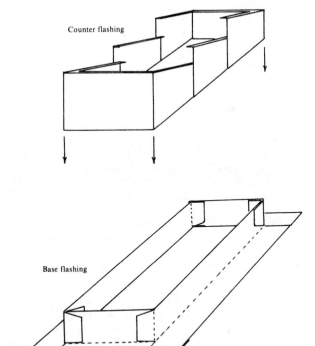

Figure 12-5   To waterproof around the chimney you will need a base unit and some counter flashing.

Install the base flashing before the chimney passes through the roof. The flashing is then a guide (almost a form) for the masonry. Base flashing is installed at the same time, and in the same way as the shingles. Treat the four pieces of base flashing as if they were four odd-shaped shingles. That is, overlap the pieces with one another, and with the shingles.

First, shingle from the eave up to the chimney hole. Then install the bottom piece of base flashing so that its flat skirt overlaps the top of the shingles. Next, add the side pieces of base flashing; the bottom ends of these pieces overlap the first piece of flashing. Finally, fit the top section of base flashing; this piece overlaps the side pieces. Nail the flashing to the roof and daub the nail holes with asphalt roof patching compound.

Now go back to the regular shingles. At the sides and the top of the chimney space, the regular shingles lie on top of the flashing. Cut them to fit right to the bend of the flashing, coat the flat skirts of flashing with patching compound, and press the shingles down into this adhesive.

Don't make the counter flashing until the stepped courses of masonry are inside the base flashing. The tops of the counter flashing must be buried in the mortar joints, and with stone masonry you don't know how high these joints will be until you get there. Actually, if you're not adverse to cheating on the stone, this is a good place to sneak in a course of concrete blocks. The blocks have nice straight tops and evenly staggered joints. Hidden behind the flashing, they'll never be seen. I'll never tell.

Counter flashing, like the base flashing, is "shingled". Start at the bottom and overlap the lower pieces with the ones above. Use a nail to punch rough holes in the top tabs. The burr helps hold the tab in the mortar joint.

The bottom edges of counter flashing may flap loose, especially the side pieces. Fix the flappers after all the mortar has set. Raise the loose edges and swab some patching compound behind them. Press the flaps back into place. You may have to hold them down with boards and clamps until the adhesive sets.

Super-Hibachi needed an extra bit of flashing where the cupola met the chimney. The same principles apply: overlap flashing and singles so water can't run under an edge, bury the top tab of counter flashing in a mortar joint, and stick any loose bits down with the black goo. (Photo 12-13)

Once past the flashing, the chimney rises as before: set a tile, plumb it, clad it with stone. When the last tile is set, pick some stones of uniform thickness for the rain cap. The rain cap is a stone course like the others, but extending 2 inches beyond the face of the masonry below. The rain cap is meant to keep water from running down the sides of the chimney. It also seals the gap between the tile and the stone sheathing.

Plan the cap course to finish 8 to 10 inches below the top of the last tile. There should be at least 6 inches of bare tile sticking up at the top, and a few extra inches for slope. When the capstones are in place, mix a small batch of stiff concrete and work it into the gap, around the tile and between the stones. Mound it around the tile and smooth the surface in a gentle slope from the tile to the edge of the rain cap.

Let the masonry cure for several weeks before you light the first big fire.

Alternatives

Once the backyard builder understands how to solve the problems that heat can cause in masonry, endless variations and improvements on the theme of Super-Hibachi are possible. Here are a few ideas.

Photo 12-13   The same principles of setting flashing as on your house apply to the Super Hibachi roof.

For a few extra dollars, you could add a damper to the top of the chimney. This adds another means of controlling the draft, which is particularly useful when smoking at low temperatures. We use a metal cover on the chimney, but it requires an occasional trip to the roof for adjustments. Better to have a proper damper with a chain or handle that can be operated from the ground.

The oven was a whim that we rarely use. If we had not had the door and the big firebricks at hand, we probably would not have included the oven at all. The firebox could be capped with a brick dome, then covered with a work surface at counter-top height.

In the beginning, our biggest concern were those alternatives of purpose: What to do with a thing this size if the doctor ever tells us to give up bacon and maple syrup? This is why the smoke chamber was built to such extravagant dimensions. If we lose our taste for bacon, the smoke chamber is just the right size for a garden shed, or even a sauna.

The garden shed conversion is the simplest. Just seal the smoke inlet and presto, it's a garden shed. The firebox still has the main passage to the chimney and thus remains operable. It just won't fill the smokehouse with smoke. It will still heat the oven, or burn the garbage if you like. And the barbeque, of course, is unaffected. You can have the garden shed *and* the barbeque at the same time.

The sauna was our fallback plan in case the smoker didn't smoke. Beside the firebox, in a section of wall now hidden by a stack of hickory wood, I left a 7 inch hole in the wall. Making the hole is easy enough — just mortar in a piece of stovepipe, being careful to bridge the top with a larger stone lintel or a simple arch. If the smoker wouldn't smoke, we would use the hole to carry a pipe the other way, with a sauna stove inside the chamber and the smoke going out instead of in. If the sauna were to be a permanent arrangement, we would convert the smoke inlet door to a smoke exit door, and connect the sauna stove to the main chimney. The chamber would have to be lined and benches built, but the building would serve the purpose.

So now you know the secret — confidence faltered on the very brink of the backyard masterpiece. Happily, the smoker fumed exactly the way it was meant to do, and the secret hole was plugged and hidden behind the woodpile. But if your doctor makes you swear off bacon, you know what to do.

Chapter 13

Working With Stone

The first consideration for many would-be stone-builders is where to get the stone. Chances are that it isn't very far away.

It may already be in your own backyard: rock that was excavated to make room for the basement, blasted out for a power pole, debris from road building or pipe laying. Unwanted stone may have been pushed to the back of the lot while the house was being built. It lies there under the leaves and hedges, unnoticed until you try to dig a hole.

If you have to look farther afield than that, the possibilities are endless. You may need a small truck or a sturdy car to haul it home, but there's lots of stone to be had for the effort of picking it up. Here are just a few suggestions:

Fieldstone is just that — an uncut stone that's found lying on the surface of the ground. It might even be in a field. Weathering has probably rounded the corners and worn away the soft spots. If it isn't good enough, dig a little deeper. Just under the surface, you should be able to find stones from the same bedrock source; and these will be unweathered — the faces may be straighter, the corners square.

Fencerows are repositories for all those fieldstones that got in the farmer's way. The usual routine was to gather up the stones at ploughing time and dump them at the edge of the field. Again, the rounded stones may be on top and the better shapes just beneath them. You might offer the farmer a token payment, but he'll probably be thrilled just to get rid of them. If you don't know any farmers, ask at the less prosperous looking farms (stony fields are the least profitable).

Wood lots in farming country can offer a bounty of surface rock. It was common to let the trees grow on acreage that was too rocky to plough.

145

Rock cuts, where roads slice across hillsides, expose bedrock strata and spill loose stones into ditches. Ask the local roads authority for permission.

Building sites usually involve some kind of heavy equipment, gouging away at the earth. Look for stone in the bulldozed piles. Ask the foreman if you can haul it away.

Trucking companies get paid to haul stone away from building sites. Call a few truckers (they're in the Yellow Pages) and ask if they're hauling anything usable. You'll probably have to pay a delivery fee. Do check the load at the building site — not in your driveway. If the stones are too large for you to move around by hand, the time to say "no" is during the loading and not when it's sitting on your lawn.

Demoliton crews commonly level a wrecking site by bulldozing over old cellars. Normally, there's too much labour involved in digging out the old stones. Call the companies and ask if they have any stone foundations on their books. They'll probably want to sell the stone. Offer to dig it out and haul it away yourself. That will make it considerably cheaper, and will give you a chance to pick out the choice material. Timing is key — you must get in between the wreckers and the bulldozers.

Lakeshores and streams wash the dirt away from rocky banks and bottoms. It's easily visible, and often accessible near bridges and culverts.

Quarried stone is the most expensive. The costly, part, however, is the digging up and hauling away. If you buy quarry stone through a building supply store, expect to pay a premium. If you live within the driving distance of a quarry (look in the Yellow Pages), you can save most of the cost by picking it up yourself. That might not be practical for the housebuilder, but you can get all you need for a barbeque or a gate post in a small trailer or pickup truck.

The best quarry product, cut blocks for the building industry, may be too fancy for the backyard or for the bankbook. Ask the foreman if you can pick out a load from their waste pile. This might be blasting debris, or blocks that broke in the cutting. You'll find that even the waste has lots of square edges to choose from.

Dan Maruska built his cellar entrance (Chapter 7) from quarry stone. He used a gravel quarry, and took the stone from blasting rubble before it went to the crusher.

How Much

First time stonebuilders can easily understimate the amount of material needed. Stone is an extravagant medium.

For a rough idea, calculate the volume of stone in the plan. Length × width × height, measured in feet, gives the total volume in cubic feet. Subtract any significant hollow places — the firebox of a barbeque, for example — and you have the volume of stone. Obviously, not all of that volume is stone; there is the mortar between the stones, perhaps a footing, but err on the generous side and call it stone. The volume, in cubic feet, multiplied by 150, approximates the weight of stone in pounds.

The density of stone varies, of course. 150 pounds per cubic foot is only an approximation for common sandstones and limestones. Granite weighs more than that, basalt less. But backyard building is more art than science. Leave precision to the carpenters and stick with 150 pounds per cubic foot as a handy rule of thumb.

The real wildcard in estimating stone requirements is not the density of a particular species of rock, but its suitability for building. Sedimentary rock, with its flat faces and easily squared edges, may be entirely usable. Every rock in the pile will fit somewhere. But building with granite boulders involves a geat deal more waste. If you can't cut the rock, you'll waste all the ones that are hard to fit. If you do persevere with the chisel and *make* them fit, you will waste all the excess cut away from the blocks.

If you're a novice with a chisel, start with the flattest, squarest stones you can find. Even then, add 25% to the calculated volume. Let's say you want to build a gatepost, two feet square and four feet high (including the underground portion). 2 × 2 × 4 makes 16 cubic feet. Add 25% for waste and you have 20 cubic feet. At 150 pounds per cubic foot, that's 3000 pounds of rock.

If you're bringing stone home in the back of the car, you probably won't bother to weigh it. But if you're having rock delivered, you'll have to estimate the needs by weight or by volume. Truckers sometimes deal in cubic yards rather than in cubic feet. Divide your estimated volume by 27 to covert cubic feet to cubic yards.

In most cases, it's better to order too much stone than too little. The most expensive part, perhaps the only expense, will be haulage. Half a truckload may cost as much as a full truckload. More important, the excess gives you a wider choice of stones, which means easier fitting and less cutting. I won't even mention the fact that the leftovers will also be the inspiration for the next project.

Quality

Don't be confused by vague references to "good" stones and "bad" stones. This is not a moral judgement, nor even an aesthetic one. A good stone is one that is easily fitted into the structure. As I get older, the term also begins to include those which don't have to be carried far.

In general, choose stones for their suitability to the structure. They should be sound — that is, neither cracked nor crumbly. And include an assortment of sizes, all small enough to lift. Within those two limits of soundness and size, a good building stone is one with two flat sides. In a wall, one flat side is the top of the stone and the other is the bottom. The exposed "face" of the stone is actually the edge of a slab like this.

The common sedimentary stones, limestone and sandstone, break naturally along parallel planes. These planes — called the "grain" — can sometimes be seen as a pattern of faint lines along the side of the rock. Each line was once a separate layer of sediment on the bottom of the sea. Some layers bonded better than others. So, as the young sea grew up and became an old rock, it retained these strong and weak strata. When it breaks, it tends to break in fairly uniform slabs.

Such conformity is a boon to the wall builder. In a coursed structure, the fitting goes faster if the stones within each course are a uniform thickness, or simple multiples of that dimension. Each course begins on a flat base, and ends with a flat top. Thus, if you've started a six inch course, a good stone is another one six inches thick. It will fit almost anywhere. Here, a seven inch stone would be a poor choice no matter how pretty it might be.

The easiest way to match stones is to take them from the same place. They were formed at the same time, in the same way. Their weak strata are the same.

With flat tops and bottoms, and with some uniformity in thickness, coursing a stone wall is as easy as stacking blocks. Any straight edge can be exposed at the face. If there is no straight edge, you can make one with a single blow of the hammer — it can be any shape it wants.

In a veneer wall, where the stones are set "on edge", quality takes a different dimension. Now the sedimentary plane is the face. Thickness of the slab is again important because the veneer is only one stone thick, and that demands uniformity. A stone that is thinner than the rest weakens the wall. A stone that is thicker than the rest sticks out beyond the face.

In veneer, the important quality is the quality of the edge. Many of the edges may have to be cut. They will have to be cut along parallel straight lines if you intend to lay the wall in regular courses. Or, the edges will have to be cut to match the angles of the other stones if you're fitting the edges mosaic fashion. And the edges will almost certainly have to be cut to make them perpendicular to the face. A rounded edge means the stone must be balanced on a narrow point, propped up with mortar. You want a square edge so the stone can sit solidly upright ... without assistance.

Making square edges is not difficult, as long as the stone is solid enough to break cleanly. If there is a weak stratum in the middle of the stone, however, it may break in steps. A straight cut, instead of going clear through the stone, veers off along the weak plane. The only way to tell is to attempt a few cuts. If you ruin the first few rocks, it might not be your fault. It might be poor quality stone.

Where the structure is primarily a surface — a porch floor or patio, for example — the only important criterion is one good face. A round backside can be buried in the forgiving bed of sand. Rounded edges can be chipped away. If they don't break cleanly, it doesn't matter because the stone rests on its backside, not on its edge. As long as the face fits, the rest of the shape is irrelevant. Quality, for a floor, refers to the flatness of the one good face.

Heated structures, such as barbeques, are little more than mini-walls so far as the stone is concerned. Some advise against using river stones with fire, on the premise that these contain more water and the heat will cause the stone to explode from trapped steam. Quite likely. But it's also true that heat can crack just about any stone. Stone can suck up water like a sponge. If you leave your barbeque outdoors there will be moisture in the stones — no matter where you got them.

If there happens to be one in the neighborhood, you can get quality barbeque rock from a volcano. Everything else should be protected with firebrick.

Finally, quality can include aesthetics. But you might want to adjust your perspective. A rock that looks good on its own may not add much to the beauty of a stone structure. The colour, or texture, of an individual stone can get lost in the mass of the structure. What matters, to my eye anyway, is the harmony of the mass. The various sizes and colours have to mix easily. Individual beauties can make the wall look motley rather than beautiful.

Concentrate on the structure and the appearance of the stone takes care of itself. If you take all the stone from a common source, it will already be reasonably matched by size, colour and shape. The stones will be siblings from the same primordial bed. Harmony is assured. A good stone is one that fits easily into the wall.

Tools

Stone tools are simple and sturdy. Most of what you need may already be at hand.

The toughest task may be moving the stone around. The basic tools include: wheelbarrow, large pry bar, rollers, ramp, and truck (or trailer). You can rent a truck, or borrow one, or even have the stone delivered. But the other tools are invaluable for heavy stones.

A sturdy barrow is the very best means of moving stones from rockpile to worksite. Load the barrow by tipping it down on its side. Roll the rock into the pan. Then tip the barrow upright again. You can tip up far more weight this way than you could hope to lift. The barrow also makes a handy pan for mixing mortar.

A long, steel crowbar can pry stones out of the ground, or nudge them into place. You can raise an edge while leveling floors, or push weighty rocks up an inclined ramp. Look for the heaviest steel you can find — the light bars bend.

Rollers can be fat pieces of pipe, or small tree trunks. Two rollers are a minimum for the big slabs. Three make moving even easier.

Rollers work even better on a ramp. A sturdy plank will do. Two sturdy planks do even better. Ramp heavy stones up to the wall, onto a truck, across the lawn, or onto a scaffold.

There's more on how to handle heavy stones in Chapter 7.

Stone cutting means hammers and chisels. There are specialty tools for stone, but a few familiar household tools will do the job almost as well.

A long-handled "sledge" hammer is not usually considered a cutting tool. But it is the fastest way I know to convert a big, worthless lump of stone to several smaller, useful lumps.

A short-handled "mash" hammer is my third hand. The little two or three pound head is perfect for driving chisels, breaking rocks, and persuading tight-fitting stones to get in line.

Two chisels are plenty. A wide, 3½ inch "brick" chisel cuts a straight line. A smaller "cold" chisel, one inch across the blade, can shape edges or split the larger stones. The blades dull quickly. Most of the time that doesn't matter — the chisel actually fractures the rock more than it cuts it. A dull blade just softens the blow a little. For a fine line, or a fragile rock that could break the wrong way, a sharp chisel makes a difference. I sharpen blades at least once a day on the bench grinder. Put a grinder on the list of useful, but not essential, tools.

Another useful, but not essential, tool is the mason's hammer. It sports a square hammer head on one side, and a chisel on the other. It doesn't do much that the mash hammer and cold chisel can't do, but having it all on one handle speeds the work considerably. You can hold a stone in one hand and trim it with the other. The stone is shaped and back in the wall in the time it would take to find the little chisel amongst the clutter.

To lay stone straight you'll need some string, a measuring stick, and a level. A plumb bob, chalk line and chalk would be useful on some jobs, but you can improvise without them.

The measuring stick will take a beating, especially if you're using mortar. Don't ruin a retractable steel tape measure by taking it within sight of sand and mortar. A plain, old wooden yardstick is all you need. You can get it dirty, wet, even leave it out all night, and it never gets stuck in the reel.

A long level is better than a short one. The length averages out the bumps and dips on the stone. But if you already have a short level, put it on a long straight board and it's almost as good as the expensive model.

The most important equipment is that which protects the mason. Use safety glasses for stone cutting, rubber gloves for the mortar, steel-toed boots when you're moving big stones around, a dust mask for grinding and sawing. Improvise on the other tools, but don't short change your eyes and toes.

Cutting Stone

Stonecutting suffers from an overblown reputation. Mystique has elevated a simple skill to something approaching witchcraft. Unfair! The Mona Lisa may be high art, but that fact doesn't deter the handyman from painting his own garage. Likewise, carving gargoyles and fancy cornices are beyond the ken of the backyard mason, but that doesn't mean we can't cut a corner stone with amateur whacks of the hammer. It's slow work, and we're bound to turn a few would-be corners into gravel. But there's no more mystery to this than painting the garage or pounding nails. I bend the odd nail, drip paint, and break a few rocks the wrong way.

"Breaking" is actually a better word than "cutting". Hammer and chisel send the force of a blow through the stone. The hammer provides the force. The chisel concentrates the force along a narrow line, and directs it through the stone. If you repeat the force often enough, along the same line and in the same direction, you will eventually create fractures inside the stone. If all the little fractures line up in a common plane, or if they find an already weakened strata in the rock, the stone will crack and come apart. Promise.

The trick is to make it come apart in the right place.

With three precautions and one discipline, you can get it right most of the time.

Precaution # 1: Put the stone on the ground — on soft dirt or sand. A hard object, even a pebble, can cause problems under the stone. The chisel directs the force from one side, but the pebble may concentrate the force someplace else on the underside of the stone. The result is a break in the wrong place. Sand supports the stone uniformly and lets the chisel do all the breaking.

For the sake of comfort and convenience, I often cut on a wooden bench or scaffold. Always brush away the chips before trying this, though. And expect to ruin more stones on the scaffold than you would on the ground.

Precaution # 2: Check the stone for cracks. You can't always see these, but avoid them when you can. There is a high probability that no matter where you hit the rock it will come apart at the old crack. If you can convince yourself that the old crack is right where you intended to cut, fine. Hit it and the break will follow the crack.

It is possible to cut a stone that is already cracked. You can cut across the crack at something near a right angle. Or, if the cut must run parallel with the crack, keep it as far away as you can. You will also have to use less force on the hammer. Sharpen the chisel to concentrate the softer blows even more. And score the cut all the way around the stone. The stone may still break at the old crack, but technique can affect the odds.

Precaution # 3: Rocks are basically lazy. A break will look for the shortest path through the mass. So plan the cut to remove weak protrusions. Or keep the cut in the fat part of the rock, where it doesn't have a chance to detour through a weak short-cut.

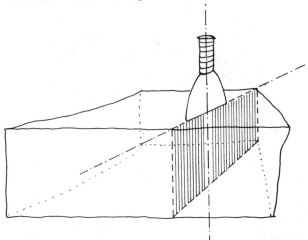

Figure 13-1   If you keep the chisel on a consistent line the break will always be true.

The *discipline* is a simple matter of remembering to keep the chisel on a consistent line. The break will run *across* the rock on a line with the blade. The break will run *through* the rock on a line with the shaft of the chisel. (Figure 13-1)

If the chisel wavers, the force will go through the rock on several different planes, dissipating the cumulative effect. The rock will get confused and won't know where to break. Rocks are notoriously slow-witted. You must give firm direction through the chisel.

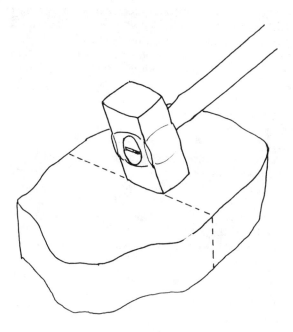

Figure 13-2   For making smooth breaks on large stones, chisel around the stone making a deep score on each surface.

Photo 13-1   Keep the chisel from wavering if you want a smooth break.

In practice, backyard building is not a precision business. If every stone were cut perfectly square, they would look like so many bricks. For our purposes, a rough break is usually close enough.

Rough cuts start with the sledge. A flat-topped slab might lack a straight edge for the face of the wall. Smack it on its flat top with the sledge. That should shatter the rock into several smaller rocks, each with a straight, unweathered edge. (Figure 13-2)

The sledge is not just a bludgeon. Look at the head. The striking face is squarish. Any edge of that square works like a big, fat chisel. The line of the striking edge determines the line of the break. If you hit the rock flat with the face of the sledge, the break might run in any direction. Strike it with the edge, and you can direct the force along a particular line.

For more precise cuts — squaring a corner, for example — use the wide chisel. Mark the cut with a yardstick or square. then chisel along the line. Remember to keep the chisel upright if you want the cut to be vertical. Work back and forth along the line, scoring a shallow groove where the mark was.

The wide chisel makes it easier to maintain a straight line. Move the chisel just an inch or so each time, over-lapping the blows so that the chisel stays in the groove.

Now tip the stone up on edge and extend the groove right around the corner. This helps to make the cut run squarely through the slab. You could even match the groove on the underside of the slab, showing the fractures where to emerge at the bottom. When the cut is well scored, increase the force of the blows, still keeping the chisel aligned with the desired plane of the break. (Photo 13-1)

Eventually, a crack should appear in the groove. The ring of the hammer will become a thud. You can

Photo 13-2   If you want a smooth surface on top of the stone, use a concrete grinder.

larger stones, and harder ones. The narrower blade focuses the force on a smaller area. A granite boulder, if it shows any grain at all, can be split like this. Align the chisel with the grain and hammer away. Stick to one spot. Eventually, you'll chip out a little niche for the chisel. The hammer forces the chisel into the niche and cracks the boulder along the grain. Now use chisels or wedges anywhere along the crack, forcing it open. With a steady hand, you can cleave the most useless sphere into two neat hemispheres. Then mark edge cuts on the flat faces of the hemispheres and go back to the wide chisel. It is slow, slow work. Never set out to build a house this way.

Dan and two burly friends spent a frustrating morning attempting to break up a table-sized chunk of limestone. It was so big that the four of us and all the levers we could muster wouldn't budge it. We either had to break it into movable pieces or plant flowers around it. Muscle and sledge hammers hadn't made a dent. So we switched to the little 1 inch chisel. I held the chisel with a pair of pliers while Dan tapped away with the sledge — little, half swings that wouldn't break my hand if he missed. In ten minutes, the chisel had made a hole for itself. In twenty, a crack appeared and the behemoth fell into four neat blocks.

At the other extreme, thin, fragile stones shatter all too easily under the chisel — usually in the wrong place. In this case, a saw is handy. Use a masonry blade in an ordinary circular saw. And don't forget the dust mask. Set the blade for a very shallow cut, and score the rock surface. In repeated passes, cut the groove a little deeper each time. It isn't necessary to saw completely through, but do make the cut the thinnest part of the stone. Break off the excess with a hammer.

You can also use the saw to improve a bumpy surface. Cut a series of parallel grooves across the bump, then chip the ridges away from the side. This process is illustrated in Chapter 2.

Other ordinary tools can be used on stone. You can bore small holes in stone with a variable speed drill. Use a special masonry bit, keep it at slow speed, and watch the temperature. Or, fit a "concrete" grinder on a flat, disc sander. Flooring specialists use these to smooth hardened concrete. I took the rough spots off the table with one. (Chapter 6)

To the stone mason, power tools are more for precision than for speed. I've used them to shape and polish marble shelves, and for fussy indoor work. But — for the backyard projects — hammers and chisels are faster, quieter, and much less dusty. With a little practice, they're almost as accurate.

force the crack, using the chisel as a wedge. Or, move along to the uncracked parts of the groove. I prefer to leave the crack and work along the groove. Forcing a crack too soon is risky if the fracture doesn't run straight through the stone.

In most cases, the first break will leave rough edges that have to be trimmed. Finish these off with the chisel.

Some bumps bulge out of the face like hillocks on a plain, leaving no abrupt edge to get the chisel into. Put the narrow chisel on these, and forget what I told you about aligning the chisel shaft with the cut. If you do that here, the chisel will just skip up over the bump. Angle the chisel into the base of the bump, as if you were going to gouge it out by the roots. The first few blows of the hammer should chip out a narrow cleft at the base of the bump. Now the chisel can get a firm bite. Straighten up the angle and hammer away.

The small chisel is even more useful for splitting

Laying Stone

Sedimentary stone can be laid on the flat in a solid wall. Or, it can be set on edge, with the flat part showing at the face of the wall.

A solid wall is commonly built in a series of horizontal layers, or "courses". Each course has a more or less level top, and the edges of the stones — the parts that show — are more horizontal than vertical. A solid wall might still have some stones set on edge, showing broad faces, but the overall impression is of parallel, horizontal lines.

The veneer wall is all faces. It is pieced together, fitting shapes to angles. The overall impression is more like a patchwork quilt than the horizontal lines of the solid wall.

A boulder wall, built of rounded stones rather than slabs, looks more like a veneer wall. It may be coursed, but the joints are too irregular for that uniform, horizontal look.

There are examples of all these styles in this book. The garden wall is solid and coursed. Dan's barbeque was built with a veneer of small, granite boulders. The gateposts are made of sandstone slabs, but some stones within it have been set on edge, veneer fashion. This is especially noticeable at the corners.

A solid wall goes up in orderly sequence. Begin at the face. If it's a mortared wall, spread a thick bed of mortar first, keeping it away from the edge. Then align a short row of stones along the face. Tap each one into the mortar as you line it up — faces flush and tops even. Keep stones of the same height together. This will make a level base on which to begin the next course.

Next, build up the back of the wall. This needn't be so neatly aligned as the front, but it must still be fairly straight to maintain a uniform thickness across the wall. Use the old-shaped lumps in the back . . . the ugly ones. But do watch the overall dimension. The top of the back row should be at the same height as the front row.

Now you have a front face and a back "face", and probably some small gaps between them. Fill the centre gaps with anything that comes to hand. Push them into the mortar bed. Slap in more mortar and more lumps until the centre gaps are filled to the same height as the rest of the course.

Fill all the joints between the stones with mortar, chopping it down into the lowest crack with the trowel. The joint at the face — the one that shows — should be filled too, but not right out to the face. The mortar is still soft, and probably won't stand up by itself in the open joint. If you try to fill it too full, too early, the mortar will likely topple out, or dribble down the face.

And don't try to push it in with your fingers — not yet. If some excess mortar does ooze out of the joint, you can slice off the excess with the trowel.

Repeat this sequence from one end of the wall to the other, and that's one course complete. It's rarely possible, or even appealing, to match the heights of the face stones along the entire course. But do remember that you're building a base for the next course up, and the more odd steps in the top of this course, the harder it will be to fit the next one.

Let's say, for example, that you're building a course about 8 inches high, and now all the 8 inch rocks have disappeared. Two 4 inch stones will do the job instead, bringing the course to a level top. (actually, you'll need something less than 4 inches after allowing for the extra mortar bed between them.)

Alternatively, you can use a much larger stone — large enough to include the likely height of the next course too. If the rock pile shows a glut of 3 or 4 inch stones, that's probably the height of the next course. So, while you're down here in the 8 inch course, stuck for another 8 inch rock, you can put in a 12 inch stone instead. The bigger stone will stick up 4 inches too high in this course, but it will match the tops in the next course.

With one course complete, slap on a new bed of mortar, and start all over again. In the second and each succeeding course, there is a new and supremely important rule: *overlap!* Each stone rests on two beneath, bridging the joint between them. That stone will be held in its place by two more above it. Gravity holds the stones down, and friction locks it together. A good wall isn't cemented together; it is locked together by this interwoven pattern.

Stand back and look at the wall. The vertical joints should be staggered, zig-zagging up the face. If there is a continuous run of joints up the face, you've given the wall a place to break.

There is an equally important overlap which won't be visible. We've bound the face together by bridging the joints. Now we have to tie the front of the wall to the back of the wall with one big stone that overlaps both faces. It spans the entire width of the wall and bonds it together. Not surprisingly, it's called a "bonding unit".

If you don't have stones large enough to span the wall, overlap two stones by at least 6 inches. One begins at the front face and extends back past the centre of the wall. And the next course includes a second slab that starts at the back "face" and comes into the wall far enough to overlap the first stone.

The number of bonding units needed depends on the design and use of the wall. As a rule of thumb, I try to make every seventh or eighth face stone a bonding unit. If the wall has a freestanding end, you'll need more bonding units there. If the wall has perfectly vertical faces, you will need more bonding units than if the were tapered towards the top. You'll need fewer bonding units over a solid footing than over an unstable earthen base.

Rounder stones make the job a little tougher. The rules are the same: span the vertical joints, and bond the wall front-to-back. But it's no longer possible to keep an even top on every course. Now each vertical joint makes a "V" in the wall. The bottom of the next stone goes in the "V". If you match the stones by size, you'll make another even row of "V's" on top. It's like stacking oranges — it works much better if they're all the same size.

The supermarket pile of oranges, however, has to be built like a pyramid, with faces that slant back towards the centre. Without the bonding units, a vertical face of matched, round stones would topple away from the stack. The tough part of building a boulder wall is bonding front-to-back. That's where the metal tie comes in.

The tie, also called a "brick tie", is a corrugated or perforated strap that keeps the face of the wall bound to the back. Their advantage is that they're so much easier to find than a cigar-shaped bonding stone for a boulder wall. You can find ties at any building supply store for pennies apiece.

When you begin to feel that it's too much trouble to build a boulder backing behind a boulder face, you might consider backing the wall with concrete blocks and bonding the face with metal ties. Now it's a veneer wall.

The veneer wall is built in a different order. The block backing, or core, is erected first. Lay one end of the ties in the mortar beds, leaving the other end free. Then add the face. When the mortar bed at the face comes near the free end of a tie, bend the tie into the mortar, and lay a stone on top.

With a veneer boulder wall, you can leave the boulders whole and let the bumps project from the face. Or, you can split the round stones and expose the split face.

A Straight Face

Ordinary walls are made straight and plumb with a variety of aids. Corner stakes and batter boards mark the perimeters. A string from corner to corner defines the straight wall. A plumb bob verifies the vertical. A level checks horizontal lines.

A masonry wall begins at the corners. Build the corners at least one course higher than the rest. Then stretch a string from corner to corner. The string marks the top of the course and keeps the face straight. Fill in the face from corner to corner, aligning the stones with the string. If the corners are plumb, the wall between them will be plumb. Finish the course, build the corners higher again, and raise the string for the next course.

That's the official way. The stonemason adds one more tool that is better than all the rest — the eyeball. The structure should not only be straight, it should look straight.

Chapter 14

Concrete and Mortar

Just so we understand one another, "cement" is the ingredient. Concrete and mortar are the products. That was not a cement sidewalk that skinned our knees; that was concrete. That's not cement between the stones; that's mortar.

Cement is a fine, dry powder that comes in bags. When you mix it with water, it begins a chemical reaction which alters it. When the reaction is complete and the water is gone, it is not dry cement, but an altogether different, rock-like substance. You can grind it into a power again, but it won't be cement. You can wet the powder again, but it won't get hard this time.

Mix cement, water and sand together to make mortar. The sand provides some strength, and the wet cement binds the bits of sand together. Put the mortar between two stones and, as it cures, the mortar will adhere to the stones and bind them together. It will also fill the spaces between the stones with a hard substance that prevents them from wiggling loose.

Even if the mortar had no cohesive value, it would still have value as a filler. A hard substance which fits perfectly in the spaces between the stones would make perfect shims in a dry stone wall. Add the cohesive factor and the result is a very strong wall.

Strength comes not so much from the mortar which has, after all, been compressed to quite thin layers in the wall. Though rock-like, a ¼ inch seam of mortar is no stronger than a ¼ inch sheet of stone — you can break either one in your hands. The strength is in the larger stones, and in the combination of stone and mortar.

Concrete is more like the stone wall than it is like mortar. Concrete is made by mixing cement, water,

sand ... (so far, that's mortar) ... and gravel. The gravel may be little stones rather than big wall stones, but the result is almost the same. The rock-like substance of sand, cement and water binds the stones together and fills the spaces between them.

Every part of that is important. If you left out the sand, the cement and water would still react and turn hard, binding the stones together. But it would be a weak bond, with lots of spaces between the stones. You could pull the end result apart like a popcorn ball.

If you remembered the sand forgot the cement, it would still be sand and gravel when it dried ... with no more staying power than a sand castle, drying in the sun.

If you leave out the water, or let it dry too soon (before the reaction is complete), the cement powder remains a powder and the sand and gravel have nothing to hold them together. Now it's a sand castle with grey powder in it.

The important thing to remember is that concrete and mortar become hard and useful because of a chemical reaction. The reaction isn't instant, but takes many days to complete. If the mixture dries before the reaction is complete, the reaction stops. The resulting substance won't be as strong as it could have been. Concrete and mortar must be given lots of time to cure before they're allowed to dry.

Ingredients

Sand and Gravel (or crushed stone) are relatively cheap when bought in bulk. Unfortunately, they are also a nuisance to store in bulk.

154

First, the savings. We pay $5 to $10 a ton here, depending on the grade and the distance delivered. For that price, it is dumped at the job site.

When the family half ton could still make it to the pit, we could fill it up with whatever we wanted for under $5 a load. Compare those prices to $6 for a paper bag of sand when it comes pre-mixed with a little cement.

Bulk is cheap, but one source requires a truck, and the other requires a place to store large heaps of the stuff.

Storage is more than having the room to pile it. Over time, the grass will grow up through the sand, the wind will blow it away, rain will wash it onto the drive and, worst of all, the neighborhood cats will find it. Storage takes a large sheet of plastic under the pile, sides to keep the pile in place, another tarp over the top, and a replanting job when the sand is gone and you discover the bare patch on the lawn.

As a long-time masonry nut, I've become inured to having piles of sand around the place. But I can still understand why Liz shudders when another dump truck pulls up the drive. How to buy is between you and your spouse and the neighborhood cats. What to buy involves simpler advice.

"Pit run" refers to sand, or sand and gravel, just as it comes from the ground. It is usually the cheapest form of aggregate. The natural proportions of sand, gravel and larger stones will vary. In other words, slways check the pit before you order. For most concrete jobs, you will want just a little more gravel than sand (about 4:3 by volume). And if there are many stones larger than 1 inch in diameter, you may have to screen the pile. Still, the price makes pit run worthwhile for most basic concrete jobs. Check the pit carefully, and tell the loader what you want it for.

"Crusher run" refers to aggregate as it comes from the crusher. Some quarries reduce large stone to crushed stone (looks like gravel with sharp edges) mechanically. The product, before screening, is a mix of large and fine particles. You can use this for concrete, but check the ratio of sand to gravel, just as you would pit run.

"Screened" aggregate has been sorted by particle size.

"Masonry sand" has had all the gravel screened out. Unless you're prepared to screen your own, ask for masonry sand for any mortar job.

"Sharp" sand is not a dated description of a pretty beach, but a reference to the quality of the particles. Some sand particles are worn and rounded (beach, sand is often like this), while others have sharp, abrasive edges. You can feel the difference by rubbing it between your fingers. Sharp sand makes stronger masonry than smooth sand.

How much to buy depends on what you plan to build and the quality of the stone. Start with an estimate of the overall volume of masonry. For concrete, the answer is easy: you'll need as much mixed aggregate as the volume of the pour. For example, a 2 cubic yard floor will take 2 cubic yards of mixed sand and gravel.

If you're buying sand and gravel separately (unmixed), remember that sand will fill spaces between the gravel stones. Buy in the ratio of 4:3, by volume. And order about 25% more, in total volume, than the volume of the pour.

For mortar, the first question is how flat is the stone. Flat, uniform shapes can be assembled with relatively narrow joints. Rounded stones take thicker joints. Experience is the only guide. Begin by assuming that for regular slabs up to 20% of the wall might be mortar. Here, 20% of the volume of the wall is an extremely loose approximation of the volume of sand required for mortar.

If you have access to pit run sand, or if you've bought good sand and it's now full of toys and cats, you may have to screen the sand. It's a simple, but time-consuming process.

A cast-off window screen will do the job, but the fine mesh slows the task considerably. Ask at the hardware store for one eighth or three sixteenths mesh. Nail the mesh on an almost upright frame. Shovel the sand towards the top of the screen. The finer particles pass through the screen and pile up behind. The gravel, or whatever, rolls down the face of the screen. Continue re-screening the gravel at the bottom until most of the sand has passed through, then toss the gravel aside. You don't have to screen the whole pile at once. Just keep ahead of the mixer. (Figure 14-1)

Figure 14-1   A simple screen can be constructed with cast off window screen and scrap lumber.

Cement comes in several forms. The basics are "portland" and "masonry". Portland makes a hard product. It's the only kind you need for concrete. Masonry cement contains more lime. The lime makes it stickier, less brittle, but weaker than portland. Masonry cement is for mortar rather than concrete. The store will have other varieties: for stucco, parging, patching, and so on. But, for stonebuilding, let's stick to portland and masonry.

Cement is sold by the bag. Buy it in small quantities, no matter how big the job. Portland, in particular, is difficult to store for any length of time. Even indoors, a bag of cement can begin to harden from the humidity in the air. Lumpy cement is useless. Even dealers can have storage problems. Load your own at the store. If the bag feels stiff and hard, put it back and try another.

Bags of pre-mixed ingredients are now available. Sand and cement for mortar, or gravel mixes for concrete. It's handier than bulk, and doubtless worthwhile for small jobs, but it is expensive.

For concrete, the alternative to mixing your own is the ready mix truck. You can order a truckload, or pay a little more per yard and buy metered concrete. The metered trucks mix the ingredients at the site, and you pay only for the amount you use. Most large concrete jobs — footings or floors, for example — should be poured in one operation. Spacing many small batches over several days makes finishing a headache, and it weakens the concrete. For the big jobs, ready mix is the only answer.

Tools

The common backyard mixer is an expensive tool for the casual user. And it really is for concrete rather than mortar. Don't rush out and buy one for a single project. Start small and mix by hand. Or rent a mixer for the first few projects.

A concrete mixer has fixed vanes attached to the inner sides of the drum. A mortar mixer has moving vanes inside a fixed drum. The mortar mixer may be three times the price of the concrete mixer. It is possible to make mortar in a concrete mixer, but only by adjusting the tilt of the drum (until the bottom is almost vertical), and by mixing smaller quantities.

Hand mixing requires a pan with a flat bottom and sloped sides. You can make one from plywood, but a wheelbarrow with a steel pan is better. Being able to wheel the mix from sandpile to job site to clean-up spot and back again is a plus.

Mixing also requires a measuring scoop, water, and a hoe. You could measure the sand and cement with a shovel, but a smaller scoop is easier to fit in the cement bag and makes less mess. Unwind the garden hose and use it for clean-up as well as mixing.

Big buckets will hold water from trowels and for dunking the stones. Keep a wire brush beside the bucket to clean the stones before you dunk them.

Trowels are the most elemental of tools. But the variety of shapes and sizes can be a little overwhelming.

A large, triangular trowel is the most useful, all-around tool. You can scoop up mortar and slap it on the wall, slice mortar into the joints, cut off the excess, bang down rock with the butt of the handle. The best, sturdiest trowels can even trim a stone.

Small, triangular trowels are handy for finishing in hard-to-reach corners, and for little else.

The large, flat, rectangular trowels are for finishing surfaces in the final smoothing operation. Unless you're finishing a concrete floor, you can live without this one too.

A small, rectangular blade makes flush pointing (finishing joints in a flagstone floor, for example) easy. It doesn't have to be an official trowel, though. I've flush pointed with a butter knife.

The skinny "S"-shaped thing is a pointing trowel, for fancy joints in a wall. Buy one if you want a formal finish, but you can joint quite adequately with a finger.

Those are the basic shapes. Within each style, you will find a score of different sizes and grades. Pick the size that feels well-balanced in your hand. And pick the best quality that the budget will allow. The best are made of heavy steel, and have a solid connection between the handle and the blade.

Mixing

Recipes for concrete and mortar are as varied and personal as any cook's preference. If you remember these basics, then you can tailor the mix to meet your own needs:

1. *The basic ratio of sand to cement is 3:1* (by volume). Three scoops of sand should be mixed with one scoop of cement. Mix twelve scoops of sand with four of cement. Nine with three. Three times as much sand as cement.

2. *Portland cement makes a hard mix; masonry cement makes it sticky.* For binding bricks or blocks in an ordinary wall, a straight masonry mix is fine. For concrete, use a straight portland mix. Stone masonry is between those extremes. The fat joints warrant the strength of portland, and perhaps some coarser aggregate in the sand. But we'll want to keep the binding properties of masonry cement. For most stone work, I compromise by using two parts masonry to one part portland.

3. *Sand and gravel won't add, cement will.* Mixing masonry cement with portland cement does not change the overall ratio of sand to cement. It is still three to one. The "one", however, is now the sum of two cement ingredients. Now a complete mix might be nine scoops of sand, two of masonry cement, and one of portland. Two plus one makes three scoops of cement in total; and nine to three is the right proportion.

Sand and gravel, however, cannot be added like that. Mix one scoop of sand with one of gravel and the total volume will be less than two scoops. It's impossible to generalize on exactly how much of the sand fills spaces in the gravel. It depends on the relative sizes and shapes of the particles. Just remember, when mixing concrete aggregates, that the whole is less than the sum of the parts (by volume).

The table below is nothing more than a summary of personal preferences. And, if you follow the guides, you can amend these recipes to suit yourself.

	Concrete	Concrete	Mortar for blocks	Mortar for stone
Gravel	4			
Pit Run (sand & gravel)		6		
Sand	3		3	9
Masonry cement			1	2
Portland cement	1	1		1

No, I haven't forgotten the water. Water isn't a measured ingredient like the others. The moisture in the sand, the heat, even the humidity in the air affect the amount of water that must be added. The best approach is to add the water in small quantities until the mix is the right consistency.

For small jobs, mix the mortar in a barrow. (Photo 14-1) First the sand, and then the cement. Stir the dry ingredients with the hoe. Start slowly — too much enthusiasm will send away clouds of cement on the wind. When it's all a uniform grey, make a shallow crater in the middle and pour in some water. Pull the sides of the crater down into the pool and hoe the mud until the water is gone. There should still some dry material at the edges. Pull the edges into the middle, make another cratar, and add more water.

As more and more of the dry material turns to mud, add the water in smaller doses. The last addition is a tiny trickle. At end, every trace of dry material has been scraped from the bottom of the barrow and turned to crumbly mud. It might still appear to be just a drop on the dry side of the smooth, peanut-buttery texture that makes good mortar. But the final mixing

Photo 14-1   For small jobs, mix the mortar in a wheel barrow.

should remedy that. If more mixing won't smooth it out, add another dribble of water and try again.

The usual — no, the inevitable — beginner's mistake is to think that the mix is far from done and add just a little more water than it really needs. That last splash turns a perfect mix into a soupy glop that won't stand up in a joint. don't try to dry it out. And don't try to use it. It will only run down the face of the stones and spoil the appearance. Throw it away and start again. It's like learning to let out a clutch — a lot at first, then a teeny bit at a time; and it has to be done by "feel".

When it slides off the hoe in creamy slabs that try to keep some of their shape, the mortar is done. Smooth the top of the pile and cover it against the sun and the wind. If it's well-covered, you have about two hours to use it up. Less than that is better.

If the mortar starts to get stiff before you've had a chance to use it all, you can sprinkle it with a little water, but it's better to freshen a stale batch by mixing it again. Turn it with the hoe, chop it, and shape it into a pile again.

An electric mixer does the hoeing for you. The hard part — knowing how much water to add and knowing when it's done — is still up to you.

To make mortar in a concrete mixer, start the drum turning and add the dry ingredients. First the sand, and then the mortar. Let them mix to a uniform colour, just as we did in the barrow. Then add the water.

A garden hose is better here than a bucket. Start with a little spray to rinse the dry stuff away from the lip of the drum. Then direct a hard spurt at the back of the drum. Don't let it run long — just one penetrating squirt at a time. The idea is to drill the water deep into the mass of dry. If the water just wets the front of the mass, the whole glob sticks to the back of the drum, turning round and around without mixing.

When the stream can penetrate far enough to wet the steel behind the glob, the mass soon begins to fall free of the back, folding over on itself as it turns. Now it is beginning to mix. It might look crumbly at first, but as soon as it starts to fall away from the back of the drum, ease off on the water. If mixing doesn't smooth out the texture, add more water in very small amounts.

When the globs slide off the mixing vanes in smooth slabs that tend to hold a shape, it's done. Put the barrow on the other side of the mixer and tip the drum to dump. Small dabs of mortar might try to stick in the drum, particularly around the vanes. Scrape them loose with a trowel.

Concrete mixes a little differently. The stones roll around and agitate the masses (which has nothing to do with politics or music). And concrete can be a lot wetter than mortar. For a single mix, start with the dry ingredients and add water just as we did for the mortar — except add more water. It's finished when the mix is porridge-like: just barely wet enough to flow, and too wet to fold.

The wetter concrete leaves more mess in the mixer. As the mess builds up, it cakes at the back of the drum. A purist might scrape it clean each time, but there is a faster way. Begin the second mix with a shot of water. Now add the gravel. The stones act as an abrasive, scouring the messy bottom and sluicing it around with the water. (You can't do this with mortar, which lacks the stones for heavy scouring). Measure in the sand and the cement, let it turn a few times, then add more water until the mix is the right consistency.

Both concrete and mortar can be over-mixed. When it gets to the right consistency, let it turn for a few more minutes, then either dump it or shut the mixer off. If you're not ready to use it yet, don't leave it turning in the drum for half an hour. Shut off the

motor and let the mix sit. When you're ready, give it a few turns, then dump it.

When you're ready for a break, hose out the drum and leave it down in the dump position. Less than fastidious cleaning will turn this useful tool into a lawn ornament in about three days. This is one of the reasons why I prefer to mix small batches of either concrete or mortar by hand. The clean-up is easier.

Clean all masonry tools before they have a chance to dry. If it won't come off with the hose, scrub tools with the water and a very coarse rag. A burlap bag is ideal. Trowels and hand tools can be cleaned by jabbing them around in the gravel.

Pouring Concrete

Concrete flows like soupy lava. It needs forms. A form may be as simple as a trench in firm earth, or as elaborate as the modular forms we used to pour an underwater base for the dock (Chapter 12). The form shapes the pour, and holds it there undisturbed until the mix has cured.

There are two things to know about forms:

First, wet concrete hates to be confined. It will leak through gaps and push out the sides of poorly-built forms. Forms don't have to be built of good lumber, but it must be sturdy. Brace it in place; and nail braces at corners and across the tops of long straight sections.

Second, wet cement is much too dumb to know the difference between a stone and a board. It will adhere to either. If you don't want the form to be part of the finished product, coat the insides with an oily film. Even a swipe with a greasy rag will later allow you to remove a form board cleanly. Or, line the form with waxed paper.

Beyond that, form building is a matter of making shapes.

Pour the concrete into the form, then move it around with a shovel. The shovel, however, should be more of a prod than a scoop. When you probe a wet mass of concrete in a steady fashion, the concrete flows. The idea is to *flow* the concrete around the form and into the corners. If you try to drag it around with, let's say, a rake, the stones separate from the sandy part and the concrete loses its consistency.

As each load is dumped into the form, prod with the shovel, even if you don't have to move it anywhere in particular. The prodding mixes the separate loads together into a single, stronger mass. And it works out the holes and air pockets.

When the form is full, screed, or strike off the excess with a long, straight board. Rest the ends of the board on the sides of the form, then "saw" the board back and forth, pusing the excess concrete ahead. Work end-to-end, then turn around and come back, side-to-side and corner-to-corner, until the humps are gone and the valleys filled.

As you screed, some water rises to the surface. When the surface water has seeped away, it's time to "float" the top. The float is a flat, wooden trowel. It can have a long handle, or a trowel handle. If you have trouble buying one, it's because many people make their own. (see Chapter 7). Glide the float back and forth across the surface. You'll have to raise the loading edge a tiny bit, just to avoid dragging up stones. And you'll need little, if any, pressure. It does "float" across the surface, riding on the skim of wet sand it brings to the top. This, in fact, is the purpose of the float — to settle the stones a little lower and create a thin, top surface of sand and cement. Without the stones, this top layer can be smoothed to an almost glassy finish.

That final finish requires you to take a break. Leave it alone until the surface is firmer.

When you can make a fingerprint in the surface, but can't push your finger in much further than that, it's time to finish the surface with a steel trowel. Use it as you did the float, but with more pressure this time. Back and forth in overlapping arcs, raising the leading edge to avoid digging in. The pressure should raise a damp skim to the surface. If it doesn't, press harder, or dampen the top with a fine mist.

Using Mortar

Add mortar to the wall with vigour. Slap it onto the stones, filling all the little gaps. Spread it like peanut butter, almost to the edge. Don't spread it right to the face or the next stone (like the top of the sandwich) will squish out the filling and make a mess.

Slice the mortar into vertical joints, packing them full to the top. Then chop the whole bed with the side of the trowel. A good mortar bed looks like a freshly ploughed field — plump and furrowed.

Now it's ready for a mosaic of stones. But first, remember that the longer we can keep the mortar wet, the stronger it will be. If we put a dry stone on the mortar, the stone will absorb some moisture. The mortar will dry where it touches the stone, leaving a weakness just where we wanted the bond. Dunk the stone in a bucket of water. The hotter and dryer the weather, the more important this step becomes.

Now plunk the stone in the mortar and tap it into place. When the course is complete, fill the joints and slap on another bed of mortar.

At the end of the day — or sooner if the day is hot or you've forgotten to wet the stones — the mortar is ready to be pointed. Pointing forces mortar back into the joints, and smooths the mortar surface. The slick surface isn't just for looks. It sheds water that might otherwise erode the joint. Pointing protects the wall.

If the mortar has squeezed out too far, slice if off with the trowel. If, on the other hand, you need additional mortar to fill the joint, put some fresh mix on a large trowel, hold it up to the joint, and push it into the crack with a narrow trowel or a finger.

Force the mortar in hard, running back and forth along the joint with a wet trowel or finger (put the finger in a glove first!). When it's packed, you can smooth it with a few more swipes of the wet trowel or finger.

The surface of the pointing is a mater of fashion. I prefer a recessed, or "raked", joint in a stone wall. It highlights the stones and is easily done with a finger. A flush joint is essential on a floor, and requires a small, flat trowel. Flush joints are also common on older stone walls, and among some European masons. Anything fancier than that requires a special pointing trowel.

Curing

The essential lesson is to keep concrete and mortar damp for as long as you can. The first step is to shade and cover any fresh work. Keep the surface from drying out by slowing the evaporation. Damp burlap is good; damp burlap under a sheet of plastic is even better.

After 24 hours, the surface should be hard enough to stand a gentle hosing without erosion. Keep it on a fine spray, or use a watering can. After three days, the surface is hard enough to shed more water than it soaks in.

Curing continues for weeks, using up the water inside the concrete or mortar. The structure just keeps getting harder. This needn't delay you. Avoid over-stressing any fresh work. Don't, in other words, break rocks on the new concrete floor, or chisel at a stone that was laid in the mortar only yesterday. But you can walk on concrete as soon as it's hard enough that you're not making footprints.

Covering the work with plastic also protects it from the other extreme: rain. A shower makes a mess of a freshly laid wall. And a suden frost is as bad. Cover up.